Mechanical Engineering
Level 2 NVQ

Mechanical Engineering Level 2 NVQ

David Salmon

Penny Powdrill

Newnes

OXFORD   AUCKLAND   BOSTON   JOHANNESBURG   MELBOURNE   NEW DELHI

Newnes
An imprint of Butterworth-Heinemann
Linacre House, Jordan Hill, Oxford OX2 8DP
225 Wildwood Avenue, Woburn, MA 01801-2041
A division of Reed Educational and Professional Publishing Ltd

A member of the Reed Elsevier plc group

First published 2002

British Library Cataloguing in Publication Data
A catalogue record for this book is available from the British Library

ISBN 0 7506 54066

Composition by Laserwords Private Limited, Chennai, India
Printed and bound in Great Britain

PLANT A TREE
BTCV
British Trust for
Conservation Volunteers

FOR EVERY TITLE THAT WE PUBLISH, BUTTERWORTH-HEINEMANN
WILL PAY FOR BTCV TO PLANT AND CARE FOR A TREE.

Contents

Preface

Mechanical Engineering: Level 2 NVQ has been published in response to the new Performing Engineering Operations occupational standards. The book has been written with the needs of modern apprentices, trainees and college students in mind and is designed to update and extend David Salmon's books, *Engineering: Mandatory Units* and *Engineering: Mechanical Units*.

Mechanical Engineering: Level 2 NVQ takes an interactive, student-centred approach and contains large numbers of training exercises designed to be used with portfolio building in mind. By working through the exercises the trainee will be able to demonstrate both underpinning knowledge and practical ability. There is space provided for witness testimony, and trainees are also encouraged to produce computer-generated material for use as evidence for Information Technology Key Skills.

The book is based around the EMTA standards for Performing Engineering Operations, incorporating the three mandatory units, two support units and three optional units with the objective of providing a solid background of knowledge and skills in mechanical engineering. The completed book will provide evidence for Level 2 NVQs awarded by EAL (EMTA Awards Ltd) or City & Guilds (scheme 2251).

Additionally, all six Key Skills are addressed with assignments designed to allow a candidate to acquire expertise at Level 2.

The authors hope that the relatively informal style has produced a text that is user-friendly and one that encourages its readers to enjoy their training.

David Salmon
Penny Powdrill

Acknowledgements

The authors and publisher are grateful to the following companies for their help during the preparation of this text, and for permission to reproduce copyright materials:

Delphi Lockheed
British Motor Heritage Trust
Britool Limited
Cincinnati Milacron
Danaher Tool Group
Domer Tools
Ford Motor Company
Global Supplies Catalogue
Hattersley Newman Hender Ltd
HSE Books
Hydra Tools International

Kennametal Hertel EDG Ltd
L.S. Starrett company
Mitutoyo UK Ltd
Moore & Wright Ltd
Nelsa Machine Guarding
  Systems Ltd
Pratt Burnerd International
Rubert & Co
Silvaflame Co. Ltd
The 600 Group
WDS Standard Parts

Thanks also to Mr. R.L. Timings for allowing the inclusion of some illustrations from his engineering books which are published by Pearson Education, and the Wyedean Insurance Company.

Chapter/Unit reference

	Unit title	Main evidence source	Supplementary evidence source
1	Working safely in an engineering environment	Chapter 1	2, 6, 7, 8, 9, 10, 11,
2	Developing yourself and working with other people on engineering activities	Chapter 2	Planning, & training records, 1, 4, 11
3	Using and communicating technical information	Chapter 3, 4	6, 7, 8, 9, 10, 11
4	Identifying and selecting engineering materials	Chapter 5	8, 9, 10,
5	Marking out for engineering activities	Chapter 7	8, 10
11	Fitting using hand skills	Chapter 8, 6	7
12	Machining engineering materials	Chapters 6, 9, 10	7
10	Using computer software packages to assist with engineering activities	Chapter 11	4
	Application of number assignments	Chapter 1, 2	7, 8, 9, 10
	Information technology	Chapter 11	4
	Communications	Chapter 2, 4	11
	Personal skills: working with others	Chapter 2	1, 4, 11
	Personal skills: improving own learning and performance	Chapter 2	Unit planning training records
	Problem solving	Chapter 12	Incidental evidence throughout

Note Evidence for '*Producing Engineering Drawing using CAD*' can be generated systematically if Chapter 3 is completed on a CAD system and evidence from Chapter 11 is adopted.

Introduction for Candidates

This book is designed to help you achieve an NVQ Level 2 in Performing Engineering Operations (PEO) and Key Skills at Level 2. NVQ stands for National Vocational Qualification. Performing Engineering Operations is based on Standards designed by the Engineering Industry Lead Body.

The standards which make up the requirements of your award are grouped into Units, and each Unit is subdivided into Elements. Most Units consists of 2 Elements but there may be 3 or more.

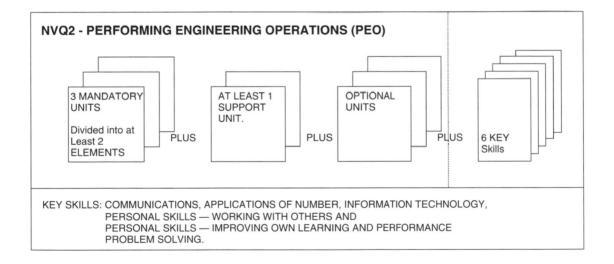

The minimum requirement for an NVQ Level 2 award in PEO is 6 Units. However, Advanced Modern Apprentices and Foundation Modern Apprentices need to achieve 8 Engineering Units and Key Skills at Level 2 as shown. AMA's then proceed to NVQ at Level 3.

NVQs are evidence-based qualifications. This means that you need to demonstrate that you can do various things as outlined in the Standards and that you understand what you are doing. When you have demonstrated that you can perform a skill to the required standard and that you understand what you are doing, you are 'signed off'. If you fail to perform the skill to the required standard – as often happens when you are learning something new – then you are given time to practise before you are assessed again.

Assessment occurs when you are ready and there may be a number of people who can assess you. Your supervisor will tell you about all the people who can act as your assessor. NVQs are not time-based and different trainees may work through the Units at different rates. There are no external exams for the Engineering award. There are external tests for three of the Key Skills.

Introduction for Trainers

With the pressure on college lecturers and the relative inexperience of some industrial trainers new to this type of work in mind, the book was devised to afford minimum inconvenience in the setting and running of a study programme for engineering trainees. However, it must be recognised that practical skills can only be learned from high quality instruction from suitably qualified teachers who demonstrate the procedures, and who can advise and supervise trainees to ensure that safe practices are observed. Trainees should follow up this instruction with plenty of practice.

A qualified assessor should approve each exercise on its completion. When the book is satisfactorily completed a qualification can be applied for.

In addition to the mechanical engineering exercises there are opportunities to demonstrate Key Skills and the book includes a number of exercises specially designed to cover the requirements of Key Skills at Level 2.

Included at the end of the book are appendices which we hope will be useful throughout a trainee's engineering career.

David Salmon and Penny Powdrill
2001

1 Working safely in an engineering environment

Exercise checklist

(Ask your assessor to initial the lower box for each exercise when you have completed it.)

Exercise no.	1.1	1.2	1.3	1.4	1.5	1.6	1.7	1.8	1.9	1.10	1.11	1.12	1.13
Initials													
Date													

Statutory regulations and organisational requirements

Safety is one of the most important elements in the working life of all people. Mechanical engineers work in an environment which can potentially be very dangerous. Only when dangers are recognised and understood can appropriate measures be taken to protect against personal accidental injury, ill health or damage to equipment. Accidents cost British industry an estimated £4.5 billion each year.

Duties and obligations under the Health and Safety at Work Act, 1974

Over the years, governments have passed safety laws to ensure that both employers and employees observe health and safety measures whilst at work. The most important of these safety laws is the Health and Safety at Work Act, which was made law in Great Britain in 1974. The Health and Safety at Work Act applies to virtually all persons at work in any job. The law places the responsibility for safe working practices onto all of the following parties:

- Employers.
- Persons concerned with premises.
- Persons in control of harmful emissions.
- Designers, manufacturers and suppliers of goods and materials.
- Employees and self-employed.

The details of the Health and Safety at Work Act are long and complicated, but the most important sections are Section 2 (General duties of employers to their employees) and Section 7 (General duties of employees at work). The content of these two important sections is explained below.

Section 2: General duties of employers to their employees

The Act states that 'It shall be the duty of every employer to ensure, so far as is reasonably practicable, the health, safety and welfare at work of all his employees'.

This extends to:

- Providing and maintaining safe plant and systems of work. All machinery and equipment should be in good working order and be safe. Protective clothing and appropriate safety equipment should be made available free of charge for employees to use.
- Arranging the safe use, handling, storage and transport of articles and substances. Included in the transport are all cranes, trolleys and appropriate routes to be used.
- Providing information, instruction, training and supervision to ensure the health and safety of all employees. This includes the provision of safety signs and warnings.
- Maintaining a safe place of work including both access and exit routes. Buildings and workplaces must comply with the safety standards and correct emergency procedures implemented.
- Providing and maintaining safe facilities and arrangements for employees' welfare at work. For example proper heating, lighting and adequate wash and cloakroom facilities.

In addition to the responsibility of the health and safety of his employees, the employer's duties under the Act apply to sub-contractors, visitors and the general public whose health and safety may be affected by his activities (e.g. harmful emissions to the atmosphere or hazards in a reception area).

There is also a requirement that every employer prepares a written health and safety policy for his employees. This policy should be kept up to date and revised as and when necessary.

A safety representative may be appointed and/or elected from amongst the employees. The representative(s) shall represent the employees in consultation with the employers. The consultation between employers and employees should aim to make and maintain arrangements to promote measures to ensure health and safety at work and to check the effectiveness of these measures.

A Health and Safety Committee should be set up if requested by the safety representatives, to review all measures taken to ensure health and safety at work and to check the effectiveness of the measures.

Section 7: General duties of employees at work

Employees are also bound by the requirements of the Health and Safety at Work Act. The rules relevant to employees state: 'It shall be the duty of every employee, whilst at work . . .'

- to take reasonable care for the health and safety of himself and others who may be affected by his actions or omissions.
- to co-operate with his employer so far as is necessary, to enable the employer to carry out his duties.

This means that all employees (this is **you**) must:

- wear suitable protective clothing provided
- use protective equipment and guards provided
- maintain their work area in a tidy manner
- behave sensibly
- apply safe working practices
- be familiar with emergency procedures
- take notice of warning and information signs
- co-operate with supervisors
- report all accidents, dangers and incidents.

Employees **must not** work on machinery without instruction and supervision, nor should they interfere with or misuse anything provided to protect the health, safety or welfare of themselves or others.

The Health and Safety at Work Act also describes the duties of two bodies, the Health and Safety Commission and the Health and Safety Executive. The Health and Safety Commission assists in the development of health and safety law and promotes health and safety through training, research and publicity. The Health and Safety Executive is concerned with enforcement of the laws. This enforcement is done through inspectors who are employed by local authorities. The inspectors have the power to inspect premises as and when they see fit: any breach of anyone's duties in health and safety laws may result in an improvement notice or a prohibition notice being served on the premises. In some cases criminal prosecutions follow.

During your training, there must be a person appointed to supervise you and offer you advice and assistance. Throughout this book, this person is referred to as your **supervisor**. In the event of your supervisor not being available, another person must be delegated to these duties and you should be aware of who this is.

You should also be aware that there is a person appointed to look after safety issues, usually called the Safety Officer. However, for information on safety issues, you would normally refer to your supervisor in the first instance. Your organisation's Safety Officer or the qualified First Aider can be approached if you need specialist advice and/or assistance about health and safety or first-aid issues.

There may be a company code of practice for additional rules in your workplace. If you follow the company code of practice and heed any warnings, you will significantly reduce your chances of being involved in an accident. Remember: you are required by law to comply with your company's organisational safety policies and procedures **at all times** whilst you are at work.

EXERCISE 1.1

State the name and location of the following persons:

	Name:	Where based:
Your supervisor:		
Your first aider:		
Your safety officer:		

Write brief comments of **your** obligations in the following circumstances:

1. You are not sure how to operate a machine required for the task in hand.

2. Your machine guard is broken.

3. You have lost your safety glasses.

4. You need to carry some hot metal parts to a cooling area.

5. The machine you are operating has spilt coolant onto the floor.

6. Your foreman tells you to clean up a mess that someone else made earlier.

7. You see a colleague struggling to carry a large and bulky load.

8. You see some jobs stacked up in front of the fire exit door.

Witness testimony

I confirm that . . . . has completed the exercise correctly.

Signed . . . . Job title . . . . Date . . . .

Sources of information relating to regulations on safety procedures

Information about health and safety is usually made available to employees by being published on notice boards and posters about the workshops. A health and safety poster must be displayed in places of work to inform employees about the framework of general health and safety duties, safety management and information about safety reps and the responsibilities of the person appointed to deal with health and safety issues. Information about the findings of safety surveys carried out in the workplace would normally be made available to employees through their safety representatives or information on notice boards.

Complying with statutory regulations

The Health and Safety at Work Act 1974 covers virtually everyone in all kinds of work. There is also legislation relating to specialist engineering processes or situations. If you require safety information about a specific task, it may be useful to refer to one or more of the following acts or regulations:

- Grinding Metals Special Regulations 1925 & 1950.
- The Factories Act 1961.

- Power Press Regulations 1965 & 1972.
- Abrasive Wheel Regulations 1970.
- Eye Shield Regulations 1974.
- Health and Safety (First Aid) Regulations 1981.
- Electricity at Work Regulations 1989.
- COSHH (Control of Substances Hazardous to Health) Regulations 1989 & 1994.
- Manual Handling Operations Regulations 1992.
- Personal Protective Equipment Regulations 1992.
- Provision and Use of Work Equipment Regulations 1992.
- Reporting of Injuries, Diseases and Dangerous Occurrences Regulations 1992.
- First Aid Regulations 1992.
- Management of Health and Safety Regulations 1992.
- Lifting and Operating Lifting Equipment Regulations 1999.

The above list is far from complete as there are many more laws related to health and safety not listed here.

EXERCISE 1.2

State six health and safety requirements in your work area which you must obey:

1.	2.
3.	4.
5.	6.

State four health and safety requirements in work areas that you regularly pass through which you must obey:

1.	2.
3.	4.

Witness testimony

I confirm that . . . . .has completed the above exercises correctly.

Signed . . . . .Job title . . . . Date . . . .

Warning signs and labels of the groups of dangerous substances

The warning signs below are designed in accordance with the classification, packaging and labelling of dangerous substances regulations. The square orange box with a black symbol is attached to labels and packaging: it gives information about any hazards related to the contents. The signs are used to mark the main groups of hazardous substances.

EXERCISE 1.3

For each of the hazardous substance warning signs below, write an example of a material displaying this warning and state the precautions which are necessary when you come into contact with it.
The first one is shown completed for your guidance.

Sign:	Example:	Precautions:
Highly flammable	Petrol.	No smoking or naked flames.
Explosive		
Poison or toxic		
Oxidising		
Harmful/ Irritant		

Radioactive RADIATION		
Environmental hazard		
Corrosive CORROSIVE		

Witness testimony

I confirm that . . . . has completed the above exercise correctly.

Signed . . . . Job title . . . . Date . . . .

Follow accident and emergency procedures

Under the Health and Safety (First Aid) Regulations, workplaces must have first-aid provision. In general, this includes a person who is trained as a First Aider, and there must also be at least one adequately stocked first-aid box that is clearly labelled. Note that first-aid boxes are primarily for the use of the registered First Aiders. Most large companies and colleges have a first-aid treatment room.

You should know who the First Aider is and where he/she can be found in an emergency. You should also know the location of the first-aid facilities.

In the event of an accident, always get the First Aider to the scene of the incident **as soon as possible**. Where possible stay with the casualty and send someone to get the First Aider. If for any reason the First Aider is not available, dial 999 and ask for an ambulance.

However, it is important to know what to do in the event of the First Aider not being immediately available and certainly in the case of electrocution.

Turning off the power before rescuing

Raise a wound after cleaning it in cold water

Rinsing chemicals off the affected area with clean cold water

Cooling a burn with cold water

If someone gets an **electric shock**, you should take the actions listed below but **on no account should** *you* **risk becoming electrocuted.**

- Switch off the electric current. **Do not touch the casualty until you have done this.** Get to know where the mains switch is.
- Send for the First Aider immediately: if you become involved the First Aider will save you too.
- Start artificial respiration if the casualty is not breathing (see below).
- Continue until further medical aid arrives.
- Reassure the casualty that help is on its way.

Allow the casualty to rest and keep him under observation.

Note: If the current can not be switched off, free the casualty from the electrical source with insulating gloves or with some dry, non-conducting material (e.g. a wooden stick or a wooden chair).

The casualty might be:

- Not breathing.
- In a state of shock.
- Suffering other injuries.
- Burnt.
- At risk of collapsing later in the day.

If someone gets a **deep cut**, the procedure would be to take the following actions:

- Send for the First Aider.
- Wash the cut area with flowing clean cold water.
- Hold the injury up high.
- Wrap or bandage firmly.
- Position the casualty in a comfortable position until further medical help arrives.
- Reassure the casualty that help is on its way.

If someone gets a **chemical burn**, the procedure would be to take the following actions:

- Send for the First Aider.
- Remove any contaminated clothing which is not stuck to the skin.
- Rinse the area thoroughly with flowing clean cold water for 10 to 15 minutes.
- Keep the affected area clean.
- Reassure the casualty that help is on its way.

If someone gets a **heat burn**, the procedure would be to take the following actions:

- Send for the First Aider.
- Rinse the area with flowing clean cold water for 10 minutes (or until the First Aider arrives).
- Keep the burn area as still as possible.
- Reassure the casualty that help is on its way.

- **Never** touch the affected area.
- **Never** apply ointments to the affected area.
- **Never** remove clothing.

Keeping a suspected broken bone still whilst the casualty is kept warm

If you suspect that someone has a **broken bone**, the procedure would be to take the following actions:

- Send for the First Aider.
- Do **not** move the casualty unless he is in a position that exposes him to further danger.
- Keep the casualty warm.
- Reassure the casualty that help is on its way.

Artificial respiration procedure

If a casualty was found not to be breathing following an accident, or after breathing fumes or smoke caused by fire the action to take would be as follows:

- Send for expert help.
- Remove obstructions from casualty's airway.
- Raise the nape of the casualty's neck and press his forehead back (to straighten the airway).
- Pinch the casualty's nose and seal your lips round his opened mouth.
- Blow firmly into the casualty's mouth until his chest rises.
- Allow the casualty's chest to fall.

Continue the process about 12 times per minute (every 5 seconds) until expert help arrives.

Nape of the casualty's neck raised and his forehead pressed back.

Blowing into the casualty's mouth with his nose sealed.

Watching his chest fall to its original position

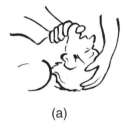

(a)

(b)

(c)

Reporting of Injuries, Diseases and Dangerous Occurrences Regulations (RIDDOR)

Following an accident a formal accident report must be made in the company's accident book. You should make sure that you are aware of the accident reporting and recording procedure for your workplace.

Dangerous occurrences that happen at work should also be formally reported. The report should be maintained in a book, a written log or a computer record. This enables measures to be taken to lessen the chances of similar accidents happening again. RIDDOR Regulations define what constitutes a major injury; a dangerous occurrence and a reportable disease.

Some possible examples for engineering situations are listed below:

Major injury

- Fractured bones (other than fingers, thumbs and toes).
- Dislocation of shoulder, hip, knee or spine.
- Loss of sight or an eye injury.
- Unconsciousness due to electric shock.
- Any injury requiring admittance to hospital for more than 24 hours.

Dangerous occurrences

- Collapse, overturning or failure of load-bearing parts of lifts or load-bearing equipment.
- Explosion, collapse or bursting of any closed vessel or pipework.
- Failure of any freight container in any of its load-bearing parts.
- Plant or equipment coming into contact with overhead power lines.
- Electrical short circuit or overload caused by fire or explosion.
- Any unintentional explosion.
- Any accidental release of a biological agent likely to cause severe human illness.
- Stop buttons or clutches not working properly.
- Collapse of scaffold over 5 metres high.

Reportable diseases

- Certain poisoning.
- Skin diseases including dermatitis, skin cancer.
- Lung diseases, occupational asthma, asbestosis.
- Infections such as hepatitis, anthrax, tetanus.
- Other conditions such as occupational cancer and hand-arm vibration syndrome.

EXERCISE 1.4

State the procedure you must follow in case of an accident to another person in your workplace.

Fill in the spaces below.

Name of the nearest qualified First Aider	Tel. no.

Name of another qualified First Aider	Tel. no.
Location of first-aid box	Nearest supply of clean water
Location of first-aid room	Person you must inform following an accident
Location of accident book	

Ask the First Aider for access to the first-aid box in your work area, and write its contents here:

The first-aid box in my work area contains the items listed below:

1.   4.   7.

2.   5.   8.

3.   6.   9.

Witness testimony

I confirm that . . . . has completed the above exercise correctly and that the trainee knows the symptoms of an electric shock, the actions to be taken before touching suspected electrocution victims and the basic resuscitation procedure.

The trainee has answered questions on the meaning of dangerous and hazardous malfunctions, and knows why these need to be reported, even if no injury occurred.

Signed . . . . Job title . . . . Date . . . .

Procedures for emergency situations and evacuation of premises

Fires can have devastating effects on industrial and domestic premises. In 1997, UK fire brigades attended 36 000 fires in workplaces, these fires killing 30 people and injured over 2600 others. After a serious fire, many firms never trade again. Fire strikes in a variety of ways and we all must be aware of how to avoid danger to ourselves and our work-mates in the event of a fire. **Fires cost lives and jobs**.

Premises can be inspected by fire officers who would award a fire safety certificate if the inspection was satisfactory.

For an outbreak of fire there must be the following three elements present:

- **Fuel** – some material that burns.
- **Heat** – the temperature at which fuels ignite (the flash point) varies from one fuel to another.
- **Oxygen** – present in the air.

If fuel, heat and oxygen are all present a fire could develop. Some common causes of fire in industrial premises are:

- Electrical faults.
- Discarded cigarette ends.
- Gas and electric welding.

Common fire precaution measures are:

- No smoking areas.
- General tidiness.
- Appropriate siting of litter bins and ash trays.

The systems below reduce the spread of fire and enable fires to be tackled before getting out of control.

- Fire alarm system.
- Fire doors.
- Fire extinguishers.
- Water sprinklers.
- Emergency exits.

The emergency evacuation procedure is not only for fires, it operates for many emergency situations. The procedure should be familiar to all personnel in the building and must be supported by information notices (in general you should close all doors and windows then leave the building by the shortest route, avoiding any flammable materials storage, assembling at a point where you will be registered as safely out of the building). There may also be maps showing the location of assembly points and how to get to them. Ideally the fire procedure should be practised at regular intervals. Get to know where the fire alarm buttons are sited, and remember what the evacuation alarm sounds like.

Use of basic fire fighting equipment

If a fire is tackled when it is small there is less chance of it getting out of control. You can tackle a small fire with a portable fire extinguisher. Care must be taken to select and use the appropriate type of extinguisher for the type of fire, as it can be dangerous to use an inappropriate appliance, particularly if there is an electrical supply in the fire. **When fighting a fire always ensure your escape route is clear**. Fire extinguishers work by either preventing the oxygen from reaching the flames or by removing the heat from the fuel; some types of fire extinguishers work in both ways. The most common types of fire extinguishers only last for about 20–45 seconds so they must be used efficiently and with care for best results. Shown below are the most common types of extinguisher, together with the types of fire for which they are most effective.

Note. BS Guidelines now recommend all new fire extinguishers are painted red (older appliances were colour coded according to their contents).

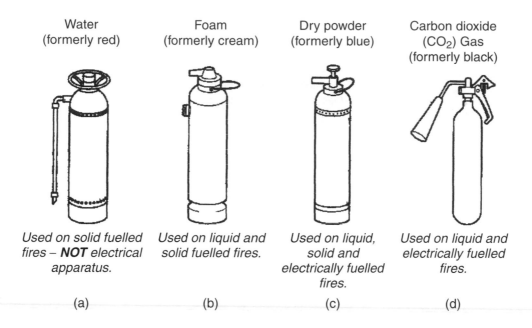

Water
(formerly red)

Foam
(formerly cream)

Dry powder
(formerly blue)

Carbon dioxide
(CO_2) Gas
(formerly black)

*Used on solid fuelled fires – **NOT** electrical apparatus.*

Used on liquid and solid fuelled fires.

Used on liquid, solid and electrically fuelled fires.

Used on liquid and electrically fuelled fires.

(a)   (b)   (c)   (d)

EXERCISE 1.5

Write here the procedure you should follow if a fire breaks out in your workplace (number 1 is completed for guidance).

1. *Sound the fire alarm to alert others to the danger.*

2.

3.

4.

5.

Common fire precaution measures are listed below. For each one, find an example in your work area and write its location in the space provided.

1. Fire doors	
2. Fire alarm system	
3. Emergency exits	
4. Fire extinguishers	
5. Water sprinklers	

In the space, draw a map of the route you must follow if you hear the emergency alarm in your work area.

State what you should do when you reach the Fire Assembly point.

State when you are authorised to leave the Fire Assembly point.

In a practice Fire Drill it takes me . . . . (time) to get to the fire assembly point.

Examine two different types of fire extinguisher and copy the operating instructions into the space below.

Type:

Instructions:

Type:

Instructions:

Witness testimony

I confirm that . . . . has completed the above exercise correctly and that the trainee knows his/her lines of reporting and limits of his/her responsibilities in emergency situations.

The trainee is aware of things s/he can do to prevent fires developing and spreading.

Signed . . . . Job title . . . . Date . . . .

Dressing prepared for work activities

Boiler suit   Lab coat

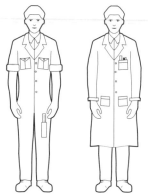

Those who work on or near moving machinery should not wear lose or torn clothing – it could become entangled in moving parts and cause serious accident. Always remove ties, rings, neck chains and other jewellery and only wear clothing that fits correctly.

Personal protective equipment (PPE) must be issued to – and used by – engineers when tasks can not reasonably be made safe in other ways. Common types of PPE include the following:

Overalls

Overalls are necessary when there are moving parts of machinery and the engineer needs to keep his loose clothing contained to prevent entanglement. Overalls also provide a means of keeping dirt and chemicals off engineer's clothes. As well as these safety uses, overalls can also provide identification and security when used as a type of uniform.

Various styles of overall are in common use.

Lightweight shoe offering no protection

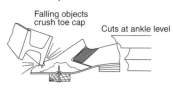

Falling objects
crush toe cap

Cuts at ankle level

Safety boots

A pair of safety boots made to BS 1870 will have many safety features. The most noticeable feature is steel toecaps. These can protect the toes should a heavy object be dropped onto the foot. Other safety features found on some industrial footwear are listed below.

An industrial safety boot

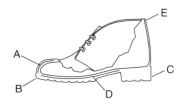

A. Steel toecap: toe protection against falling objects.
B. Chemical/oil resistant soles: for grip in oily and chemical environments.
C. Non-slip rubber sole: tread pattern improves grip on smooth surfaces.
D. Steel inner sole: Protects against penetration from underneath.
E. Ankle protection: helps to avoid injury to achilles tendon.

Safety glasses

Safety glasses

Various types and styles are available. BS 2092 lenses are suitable for protection against impact from flying debris in an engineering workshop. Safety glasses are to be worn at all times in workshops, whether you are operating a machine or not.

Safety helmet

Safety helmet (hard hat)

Wear a safety helmet when working:

- below others in case they drop something
- in places where the ceiling is low or uneven
- when a sign directs you to do so.

Hair caught in drilling machine

Hats

Hats should be worn by all when there is any risk of getting your hair tangled in a rotating tool or workpiece. The picture shows what can happen if you get the top of your head too close to a rotating drill in a drilling machine.

Ear defenders and earplugs

Ear defenders and earplugs

Necessary for those working in areas where loud noise levels (80 decibels) exist. Ear defenders look like headphones, earplugs are smaller and fit inside the ears.

Gloves

Gloves are worn for hand protection. The main types in common use are listed below.

Leather gloves – for sharp edges

Leather. Leather gloves are tough and flexible, and should be used when handling things that have sharp edges to protect hands against being cut.

Thermal insulation gloves

Thermal insulation. Gloves offering thermal insulation are made of leather and lined with an insulation material. Asbestos was used for this insulation years ago, and is now known to be a very dangerous material, but these gloves are still sometimes **wrongly** called 'asbestos gloves'.

Rubber gloves for chemical protection

Rubber (Latex). Thin rubber (latex) gloves provide protection against the effects of a wide range of chemicals, e.g. salts, detergents and harmful oils. Never use these gloves unless you are sure they will protect you against the chemical you are exposed to.

Apron worn as protection from chemicals

Apron

Rubber or polypropylene clothing provides protection against the effects of chemicals. Never use this equipment unless you are sure it will provide protection against the particular chemical you are exposed to.

Dust mask

Dust masks should be used when working in a dusty environment. If the air in a work area can not safely be filtered by fume and extraction equipment, an engineer should wear a dust mask to filter the air for his own protection. Remember that the filter only removes dust particles and not poisonous gases.

Respirator

If the air in a work area can not safely be extracted and filtered by fume and extraction equipment, then a respirator is needed. Respirators are facemasks which draw a clean supply of air from elsewhere (note the air supply pipe in the drawing). Respirators must be worn when working in poison environments or where fumes and chemicals may leak, causing discomfort or illness.

A dust mask

Other specialist protection

There are other pieces of specialist protective clothing available for those working in hazardous environments. The items of protective equipment described above are the most common, but if you are at all concerned about working in an environment, always ask your supervisor or another responsible person about protective clothing.

Personal hygiene procedures

Wash your hands at the end of each work shift and before eating or handling food. This is very important because many products in the workshop, particularly lubricants and oils, are harmful and if consumed can cause illness. Food should never be brought into an engineering workshop.

Clean your overalls regularly to reduce the chances of contaminants or irritants like oils soaking through the material and becoming in constant contact with your skin. Oil on your skin can cause a skin disease called dermatitis. Dermatitis particularly affects sensitive skin so always wash your hands before using the toilet. You can minimise the chances of getting dermatitis on your hands by applying protective barrier hand cream; this hand cream should always be applied at the start of every work session.

Respirator

EXERCISE 1.6

In each of the following three sections, complete the blank spaces as instructed:

1. *Protective clothing*
   In the drawing of an engineering trainee, it can be seen that he is dressed in a potentially dangerous way. List nine points that are a potential hazard to his safety.

a) *His long hair is not contained*	b)	c)
d)	e)	f)
g)	h)	i)

2. *Safety footwear*
   Industrial footwear has built-in safety features. Write in each space a short note to indicate the circumstances when the specified safety feature is required.

a) Non-slip rubber sole	b) Chemical resistant soles.	c) Steel inner inside the sole.
d) Ankle protection	e) Oil resistant soles	f) Quick release laces

3. *Work environment*
   Briefly explain the work environment that would require the following safety clothing and equipment, and state why they are necessary. The first one has been completed for your guidance:

a) Overalls *Necessary when loose clothing needs to be contained to prevent entanglement with moving parts of machinery.*	b) Rubber gloves

c) Leather gloves	d) Barrier cream
e) Safety glasses	f) Helmet
g) Safety boots	h) Ear protectors
i) Aprons	j) Hats

Witness testimony

I confirm that . . . . has completed the above exercise correctly, that the trainee has been observed by me continuously to be wearing appropriate personal protective equipment and suitable clothing for his/her work activities and that the candidate knows what protective clothing is available in his/her work area.

Signed . . . . Job title . . . . Date . . . .

Behaviour in the working environment

Your behaviour in an engineering workshop must be responsible and appropriate. This means that you **must always** follow these rules:

- Wear safety glasses at all times in the workshop.
- Use protective clothing and equipment.
- Know the emergency stop procedures.
- Follow instructions carefully.
- Be aware of what is going on around you.
- Use appropriate guards on all machine tools.
- Use correct lifting technique.
- Observe all signs in work areas: they are there for your benefit.
- Observe all company rules and restricted area notices.
- Ask for help and advice when unsure about anything.
- Follow all the rules laid down by your employer.

You **must never**:

- run in the workshop
- smoke in unauthorised areas
- operate a machine without authorisation
- start a machine unless you have been shown how to work it properly
- throw anything
- push persons or get involved in horseplay
- remove a guard from a machine

- distract others operating a machine by asking questions, shouting or making loud noises
- play around with compressed air lines.

It is important to develop safety awareness and be alert at all times.

EXERCISE 1.7

Give three reasons why it is always necessary to behave responsibly in an engineering workshop in order to conform to the Health and Safety at Work Act 1974.

1.

2.

3.

State where your organisation's health and safety policy is kept and who is responsible for reviewing and updating it.

Kept in: Reviewed and updated by:

Witness testimony

I certify that . . . . has been observed by me to be continuously behaving in a safe and orderly manner and observing the establishment's health and safety policies.

Signed . . . . Job title . . . . Date . . . .

Manual handling techniques

Stretching, reaching over and lifting are some of the most hazardous things you do at work. 40% of all sick days off work are due to 'handling injuries'. 54 million days are lost each year because of back problems. It is therefore particularly important to learn how to manually lift and move articles safely. You may need to use manual handling techniques if powered lifting aids (jacks, hoists and cranes) are not available and the load to be moved is within your capability.

Structure of the spine

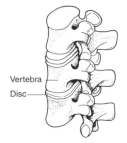

Vertebra
Disc

Once you know how to lift safely, it is important to get into the habit of doing so to prevent any injury occurring.

The human spine is made up of a column of cylindrical bones called the vertebrae. The column is held together by strong ligaments, and each disc separates a vertebra. The disc material is flexible so that movement is possible and it also allows shocks to be absorbed. If the spine is strained while lifting or carrying, the ligaments can be damaged or the discs may be displaced. Either of these injuries result in acute pain. For the back to be at its strongest, the spine should be vertical. Hence, when lifting heavy or awkward loads, you should bend at the knees and keep your head up rather than bend the back.

Back bent and head down

Wrong

(a)

Bending at the knees, head kept well up

Right

(b)

As a guide, do not attempt to lift more than 20 kg without assistance. Always ask for help with bulky or awkward loads and ensure that any load that is being transported does not obstruct your vision. If assistance is not available, use powered lifting equipment. Remember it is good practice to clear the area you are going to put the load on **before** it is picked up.

Transporting loads with manual trolleys and hand carts

Trucks are frequently used for transporting heavy or bulky articles over smooth surfaces. A large variety of trolleys, trucks and carts are sold for use in factories. Some are shown below. When using them you should maintain the speed at a minimal level, taking particular care with corners and down hills. The truck should be loaded with the biggest part of the weight low down and away from the edge. Special training and licences are necessary before you can use a powered forklift truck: their use is not covered in this book. You should neither stand nor stack articles near routes used by forklift trucks.

Sack
barrow

Pull-type
platform truck

Push-type
platform truck

Forklift
stacking truck

Housekeeping in the work area

A tidy workshop is more likely to be a safe workshop. Always keep a lookout for mislaid things that people could trip over or slip on and tidy them up – this is your duty under the Health and Safety at Work Act. Spilt oil can be cleared up with a mop and then covered with sawdust. Gangways and exit routes in particular should be clear of obstacles at all times.

Care when storing and stacking materials is essential or accidents can happen if the stack of materials is lightly knocked. If things are left sticking out of shelves, people passing by could be affected.

Take care while laying electric power cords so that they do not become 'trip wires': it is correct to position a sign to inform others of such hazards and use a protective cover over the flexible cable.

Apart from safety, general workshop tidiness is important because it makes it easy to find equipment.

Storage of tools, equipment and materials

Engineering tools and equipment should always be used correctly to minimise accidents and to keep them in good working order. Engineering tools refer to hand tools, measuring tools and powered hand tools. Equipment includes all types of equipment used in the workshop from vices and benches to lifting tackle. Materials should be stored and transported carefully and in such a way as to minimise the risk of injury.

General rules for safe use of tools and equipment

- Carefully select the correct tool for the job.
- Never use any tools/equipment without full instruction or training. If you are not sure, ask your supervisor.
- Check tools/equipment for any obvious faults, damage or excessive wear before use. Never use a faulty tool or equipment: report it to your supervisor.

- Always follow set safety rules and procedures, e.g. files must be fitted with handles, guards used on machine tools, protective clothing worn when necessary.
- For powered hand tools, use 110-volt equipment with a transformer whenever possible: the lower voltage significantly reduces the severity of electric shocks.
- Know the emergency stop procedures for all powered tools and machines.
- Take care to lay power tool cords avoiding 'trip wires'.
- Switch off the electrical supply after working electrical hand tools.
- Never operate electrically powered hand tools with wet hands, near water or flammable liquids.
- Treat all tools/equipment with respect and never use any tools or operate machinery if you are unwell or unfit to do so.

General rules for storage of tools and equipment

- All tools and equipment must be stored in a way that does not present a risk of injury to persons, e.g. sharp edges should be covered, long items should not protrude from shelves.
- After use, tools and equipment should be cleaned as instructed.
- Tools and equipment should be tidily stored in a clean and dry place.
- Tools and equipment should be stored in their specially allocated place to enable easy retrieval. If issued by the storeman, they should be returned to him.
- Accurate measuring tools may be lightly oiled before being stored to prevent corrosion.

General rules for handling and storage of materials

- Always use correct lifting techniques.
- Wear appropriate protective clothing to guard against possible injury.
- Store materials away from gangways.
- Flammable materials should be kept away from naked flames.
- Liquids should be moved only in sealed containers to avoid splashing.
- Gas cylinders must be in well-ventilated areas and away from risk of impact.
- Hot materials should be cooled in protected areas.
- Observe and obey any special instructions or safety signs.

EXERCISE 1.8

Give three reasons why it is always necessary to maintain the work area, exits and gangways in a clean and tidy condition to conform to the Health and Safety at Work Act 1974.

1.

2.

3.

Witness testimony

I certify that . . . . has been observed by me to be continuously maintaining a safe and tidy work area and that he/she has been questioned and found to be aware of the importance of safe storage of tools, equipment, materials and products.

Signed . . . . Job title . . . . Date . . . .

Disposal of waste materials

Environmental concerns are ever more important in life today. All companies have two outputs: one that everyone wants – the product – and the other that no one wants – the waste materials. Most waste materials are a hazard in the workplace: they can be in the way, they might cause fire risk and use up valuable storage space. Metal swarf and off cuts have dangerous sharp edges. Always remove your waste before it gets in your way and take it to the correct collection point. While disposing of chemicals (e.g. oil or coolant) make sure you follow the guidelines set out in the manufacturer's COSHH leaflet. Local authority sites are for domestic household products only, and industrial waste must be disposed of separately and safely. Usually most companies use contractors to deal with their industrial waste. The contract companies may recycle the waste. If you are unsure about how to dispose of any waste material you **must** seek advice.

Never mix chemicals together or put them into drains: this could be environmentally disastrous and your company will be charged under the rules of the Health and Safety at Work Act.

EXERCISE 1.9

Demonstrate to your assessor the correct technique for manually lifting a heavy load. Identify to him the alternative means of lifting or moving a load that are available to you if the load is too heavy or bulky for you to handle alone.

1. Heavy load Trainee to demonstrate knowledge and ability to safely relocate a heavy load and suggested sensible alternative of moving load.	Load moved: From: To: Alternative:

Witness testimony

I have observed . . . . move the above load in a correct and safe manner using recommended lifting techniques.

Signed . . . . Job title . . . . Date . . . .

2. Bulky load Trainees to demonstrate knowledge and ability to safely relocate a large bulky load and suggested sensible alternative of moving load.	Load moved: From: To: Alternative:

Witness testimony

I have observed . . . . moving the above load in a correct and safe manner using recommended lifting techniques.

Signed . . . . Job title . . . . Date . . . .

Control of hazards in the workplace

Hazard spotting

A hazard is something with the **potential** to cause harm. Hazards may be associated with the following:

- contaminants/irritants
- dust/fumes
- electricity
- fire
- materials handling
- material transporting
- moving parts
- pressure storage
- slippery surfaces
- tools
- toxic/volatile materials
- uneven surfaces
- unshielded processes
- working at height.

Hazard control measures

If you see a hazard, it should be eliminated, removed or isolated. Stop working when you become aware of a hazard and try to control it. The use of PPE is a last resort for hazard avoidance, so you should fit guards, use fume-extraction equipment or limit access to unauthorised personnel whenever possible.

Risk

A risk is the likelihood that a hazard will cause harm. A risk can depend on a number of different factors, e.g. the risk of a person slipping on a wet floor depends on:

- the amount of water on the floor
- the smoothness of the floor's surface
- the types of shoe sole the people have
- the number of people walking over the area
- the size of the area covered in water.

Risk assessment

Risk assessment is a method of identifying the severity of risk in the workplace. Risk assessments must be carried out by employers by law. They help to eliminate accidents and the associated cost of accidents. Risk assessments highlight training needs and help prioritise the implementation of health and safety measures.

One method of conducting a risk assessment is as follows:

- Identify and list all the **hazards** you can find.
- Identify **who** is at risk. This could be persons from the following work categories:

Office staff	Visitors	Disabled staff	Members of public
Cleaners	Operators	Site workers	Maintenance personnel
			etc.

- Assess the **likelihood** of an accident happening due to the identified hazard (you can use a scale of 1–5 as shown).

> Scale: 1 = Very unlikely to happen, causing harm
> 2 = Unlikely to happen, causing harm
> 3 = Possible to happen, causing harm
> 4 = Likely to happen, causing harm
> 5 = Very likely to happen, causing harm

- Evaluate the **severity** of a possible injury (you can use the scale of 1–5 as shown).

> Scale: 1 = Minor injury
> 2 = Major injury
> 3 = Loss of limb
> 4 = Death of an individual
> 5 = Multiple deaths

- Evaluate the **risk** by multiplying the two ratings together:

Risk = Likelihood rating × Severity rating

Following the risk-assessment process, you should

- consider situations with the highest risk 'scores' and take measures to reduce the risk by any appropriate means
- record all your findings on a risk-assessment document (see Appendix XIV).
- ensure risk assessments are reviewed regularly and revised if/when necessary.

Examples of risk assessment

Example 1. A slippery floor in a motorway service station would be rated as follows:

Hazard:	a wet floor in a motorway service area
Who is at risk:	members of public
Likelihood of accident:	4 (likely to happen, causing harm)
Severity of accident:	2 (possible major injury)
Risk assessment result	= Likelihood rating × Severity rating = (4 × 2) = 8 (out of a worst possible score of 25).

Actions:

- Put up warning sign
- Mop and dry floor
- Return frequently to see if floor is dry
- Remove sign when hazard has gone.

Review notes: floor should be inspected to see if it is dry every 20 mins.

Example 2. On a building site at the service station, bricks are stored near the edge of scaffolding platforms.

Hazard:	storing bricks near the edge of scaffolding platforms
Who is at risk:	site workers
Likelihood of accident:	3 (possible to happen, causing harm)
Severity of accident:	4 (possible death of an individual)
Risk assessment result	= Likelihood rating × Severity rating = (3 × 4) = 12 (out of a worst possible score of 25).

Actions:

- Cordon off area below scaffolding
- Put up protective 'toe board' to restrict chances of bricks falling
- Issue workers with hard hats
- Put sign up to warn of danger
- Put sign up to make hard hat mandatory
- Limit access to the high-risk area to allow access only to those needing to work in that area.

Review notes: this must be corrected by 7:45 a.m. tomorrow.

In the two cases above, there is a greater **risk** caused by the brick on the scaffolding than the slippery floor, so the risk associated with the brick must be addressed as priority issue.

EXERCISE 1.10

Part 1

Carry out simple hazard-identification inspection on both:

a) your immediate work area
b) work areas that you pass through.

Write your findings in the appropriate boxes below:

Hazards in your work area	*Hazards in work areas you pass through*
1.	1.
2.	2.
3.	3.

Part 2

Conduct **three** risk assessments by identifying different hazards and 'severity-of-accident' levels. Use copies of Appendix XIV for this part of the exercise.

Use at least **three** of the following five categories:

a) your working environment
b) tools and equipment that you use
c) materials and substances that you use
d) working practices
e) accidental breakages and spillages.

Assess the relative risks and insert suitable actions that may need to be taken.

Risk assessment 'title'	Number 1	Number 2	Number 3
Category (a–e) above			

Who should be informed on actions to be taken to deal with risks in the workplace that you are unable to deal with?	*Name of person*

State **your** organisation's procedures for controlling risks to health and safety in the space below:

Witness testimony

I confirm that . . . . asked me about the health and safety issues outlined by him/her above and that the trainee knows what constitutes a hazard and his/her responsibilities for dealing with hazards and rectifying risks in the workplace.

Signed . . . . Job title . . . . Date . . . .

Safety rules, signs and warnings

Employers put up various signs in order to ensure that employees are aware of the dangers in work areas and of the precautions necessary to reduce risk. The five categories of sign described below are the recommended shape and colour to meet with the safety signs at work regulations.

Mandatory (compulsory)	Prohibition (forbidden)	Danger warning	Safe condition	Fire equipment
Safety boots must be worn	Do not drink this water	High voltages	Emergency escape route	Fire-extinguisher point

Control measures to minimise risks

Guards

The purpose of a guard is to protect. Guards may be designed to protect people, fragile workpieces or machine parts. There are many types of guard used in engineering today. Some of the purposes of guards are listed below. It must also be noted that guards have many methods of operation. Most guards are mechanical barriers and may be transparent. Some guards are electrical and operate either as 'magic eyes' to switch off a machine if an invisible beam is broken or as interlocks to prevent a machine from working until it is safe.

Guards may be used to:

- protect operators and other persons, who may be passing, from getting into contact with moving parts
- prevent swarf and other particles flying out of machines and causing injuries
- prevent objects coming into contact with items of machinery, which may be delicate or fragile
- prevent unauthorised persons working machines
- keep people or their limbs out of danger areas
- prevent a machine damaging itself.

EXERCISE 1.11

Write in the spaces below **two** situations when you took action to ensure the working environment was made safer.

1.	
2.	

Witness testimony

I certify that . . . . was observed by me to have taken the action s/he has outlined above and that s/he knows what measures can be taken to minimise risks.

Signed . . . . Job title . . . . Date . . . .

Accident-reporting procedure

If you take care, then this significantly reduces the chances of you being involved in an accident. There is always the chance that a system could fail or another person could disregard safety rules, causing an accident to happen.

Should this happen, you need to know who to inform, how to report an accident in your workplace and where the accident book is kept.

EXERCISE 1.12

State the name of two reliable persons you can refer to on health and safety issues:

Name of person 1:	Where based:
Name of person 2:	Where based:

State an instance when you sought information about a health and safety issue from either of the above named persons. Describe the issue and the recommended solution.

Witness testimony

I confirm that . . . . asked me about the health and safety issues outlined by him/her above.

Signed . . . . Job title . . . . Date . . . .

EXERCISE 1.13

(Application of Number – Assignment 1. This exercise is specifically included to enable key skills requirements to be fulfilled.)

Task 1
Look at the following accident statistics taken from a Health and Safety Executive source.

	Construction	*Creamery*	*Transport*	*Oil platform*	*Hospital*	*All*
Number of weeks on site	18	13	13	13	13	NA
Number working on site	120	338	80	210	700	
Over 3-days lost time injuries	0	6	0	2	6	
Minor injuries	56	31	0	8	58	
Non-injury accidents	3570	889	296	252	1168	
Total accidents						

(Statistics adapted from The Costs of Accidents at Work – HSE pub.)

(a) Fill in the blank spaces on the table by adding up the rows and columns.
(b) Complete the following table by calculating the number of accidents per employee. (Answers to the nearest whole number.)

	Construction	*Creamery*	*Transport*	*Oil platform*	*Hospital*
Number on site					
Total number of injuries					

Number of accidents per employee					

(c) Complete the following table by calculating the number of accidents per week.

	Construction	*Creamery*	*Transport*	*Oil platform*	*Hospital*
Time during which accidents could occur					
Total number of non-injury accidents					
Number of non-injury accidents per week					

(d) Display the number of accidents per week for the range of industries in a form other than in a table and explain why you have chosen that format.

(e) Note that the majority of accidents did not result in anyone being injured. Calculate the ratio of non-injury accidents to injury accidents and fill in the table with your results.

	Construction	*Creamery*	*Transport*	*Oil platform*	*Hospital*
Total number of accidents resulting in an injury					

Non-injury accidents					
Ratio of injury accidents to non-injury accidents					

Express answers in their simplest form and in whole numbers

(f) Which of the above industries has the worst accident record?

Task 2

To prevent accidents due to people slipping, most engineering workshop floors are coated with a non-slip surface.

During the summer shut-down period, a machine shop measuring 30 m long × 20 m wide will have the floor recoated. Look at the following diagram:

Note there are 15 machines: six lathes, occupying a floor space of 1 m × 2 m each, and nine milling machines, each of which is 1 m × 1.5 m wide. The gangways are 1 m wide – one goes from one end of the workshop to the other and the other two gangways span the workshop. All the gangways are to be painted grey. The remainder of the floor is to be painted green.

It is **essential** that you show the stages in your working for the following calculations.

   (a) Calculate the total area to be painted grey.
   (b) Calculate the total area to be painted green.

1 litre of paint covers 10 m² and it can be bought as follows:

Size of tin in litres	Price
1	£8.50
2.5	£20.00
5	£38.50
10	£75.00

Note that orders over £500 receive a 12% discount on the total order.

   (c) Calculate the volumes of green and grey paint needed and then work out the most economical way in which it can be purchased. Calculate the final price for the paint.

The final price for the paint will be:

   (d) Convert the following from metric units to imperial (see overleaf for conversion units):

	Required unit	Answers
Width of workshop	Feet	Ft.
Width of a milling machine	Inches	Ft.
Area to be painted grey	Square yards	Sq yd.
12 litres	Pints	pt.

Approximate conversion factors:

$$1 \text{ m} = 39 \text{ inches}$$
$$1 \text{ m}^2 = 1.17 \text{ square yards}$$
$$1 \text{ litre} = 1.75 \text{ pints}$$

(e) Indicate how you have checked to make sure that your results both for the areas and for the volumes of paint needed make sense.

Task 3

In many engineering firms, heavy equipment and stores need to be moved. A small forklift stacking truck may be used for this purpose.

The truck in the stores has a clear notice that states that the maximum load is 500 kg. Look at the following information, which relates to density and dimensions of the material, and show whether they can be safely lifted by the stacking truck. Complete the table with your results.

$$\text{Density} = \frac{\textbf{Mass}}{\textbf{Volume}} \text{ kg/m}^3 \text{ (kg/cubic metre)}$$

Metal density (kg/m^3)	Dimensions	Volume (m^3)	Mass (kg)	Safe to lift?
Aluminium 2720	0.5 m × 0.5 m × 0.5 m			
Brass 8480	0.5 m × 0.5 m × 0.2 m			
Cast iron 7200	1 m × 0.5 m × 0.2 m			
Copper 8790	2 m × 0.25 m × 0.2 m			
Steel 7820	1.5 m × 0.45 m × 0.01 m			

Show your calculations here:

The above exercise was completed satisfactorily by . . . ., who could also explain how work had been checked to ensure that results made sense.

Signed . . . . Job title . . . . Date . . . .

2 Developing yourself and working with other people on engineering activities

Exercise checklist								
(Ask your supervisor to initial the lower box for each exercise when you have completed it.)								
Exercise no.	2.1	2.2	2.3	2.4	2.5	2.6	2.7	2.8
Initials								
Date								
Exercise no.	2.9	2.10	2.11	2.12	2.13	2.14	2.15	2.16
Initials								
Date								

When you start a new job, you are probably hoping to do a good job, develop yourself and obtain qualifications and promotion. There is nothing wrong with wanting the best for yourself. At the same time, most people have to work with other people, and good working relationships are an important part of the working lives of everyone. People should be able to get on with each other and work well as a

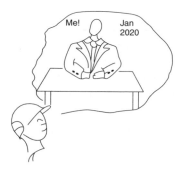

group or team to achieve the best results for their organisation and the welfare of its staff. If working relationships are good, the workforce becomes effective and efficient and working relationships improve even further.

The following exercises should assist you to help yourself and work with other people, and, in doing so, you will develop personal skills that will help you throughout your working life.

This chapter is also closely linked to the two units of key skills: **personal skills** – working with others and **personal skills** – improving one's own learning and performance. So look out for the key symbols and make sure that you fill in the evidence sheets of key skills as you proceed. You can also achieve some evidence of key skills – communications from this chapter.

In the work environment, it can be quite difficult at first to work out exactly what is expected of you, and this can cause some concern. The first section looks at some of the issues that you need to think about.

You and your organisation

In many organisations, there is a management structure that may be referred to as 'line management'. What this means is that everyone reports to someone else, except for the owner or most senior manager (i.e. the managing director or MD), and even these people probably have to account for what they do to shareholders. Usually, your line manager is your supervisor and this is the person to whom you should turn to if you are not sure of what is expected of you.

Your supervisor will explain to trainees what is expected of them as far as their actual job is concerned. It is useful to remember that a firm or a factory is in business to make money. This may seem obvious, but employees can only be paid as long as the firm makes money, and this probably involves the firm in selling goods or services at a quality that the buyer expects. Quality assurance (QA) establishes the standard of work expected, and although trainees are not expected to be able to produce perfect work from the beginning, they are expected to know what is expected of them and work toward providing the quality of work that the firm expects.

Thrust SSC – world land speed record 763.035 mph

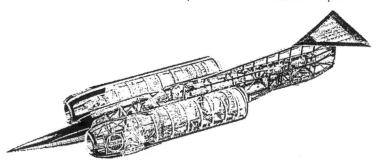

Exciting new projects in Britain, such as Thrust SSC, the Channel Tunnel and the London Eye continue to operate only if the initial standard was set at an appropriate level.

Finding out what is expected of you in terms of standards means that you must be prepared to ask questions.

Developing training plans

When you start a new job, you must be given some form of 'induction'. Depending on the size of the firm you work for, the induction you receive may be more or less formal. Whatever form the induction takes, everyone needs to know the same information. You will need to know:

- more about the firm for whom you are working, what they make or what service they supply, how large the firm is and how many other people work there
- basic 'comfort' information such as where the toilets are located, where the canteen is sited, whether smoking is permitted on site
- when and how you get paid
- the terms and conditions of your employment including the hours you work, how many days holiday you are entitled to and what you have to do if you are too ill to come to work.

All the information given in the preceding list is really basic and most of it would have been told to you before you started work, but issues are easily forgotten or may need to be clarified once you are actually employed.

How many people work here?

£ Into your bank account

WC ?

No smoking on site

Canteen first floor

Start

Finish

Holidays 22 + statutory

EXERCISE 2.1

Complete the following exercise to show that you understand what is required of you as you start your training:

State the name of the company for whom you work:	
Outline very briefly what this firm does:	
What is the name of your trainer/supervisor?	
When you have completed your training to NVQ Level 2, what engineering skills do you expect to have achieved? (What will you be able to do?)	

Witness testimony

The above exercise has been completed satisfactorily by . . . .

Signed . . . . Job title . . . . Date . . . .

Once first fears have been overcome, you will quickly appreciate that there are lots of other questions that you need answers to and this is where training plans (action plans) come in. Your firm may well have a training scheme for apprentices and trainees and you may be informed that you will be attending college once a week to learn more about the theory side of your work or to extend 'key skills'.

However, even with a training scheme, there are always things people do not understand, so it is a good idea to keep a note of questions you want answered. Jotting questions down in a diary or notebook ensures that things are not forgotten and that when you see your supervisor or training advisor you get the answers you want!

Your initial training plan is really a contract between you, your company, the Government (via some representative) and the training provider. The training plan is drawn up and this outlines the main stages in your training and may set dates by which various stages will be completed or when there will be official assessments. National vocational qualifications (NVQs) have assessment plans by which you know what stage you have reached in your training. NVQs require you to provide evidence that you are competent and can carry out a task skilfully and prove that you understand what you are doing. NVQs are not time-based, so if you show that you have a real aptitude for some areas of work, then you may find that you are passed as competent in these areas quite quickly, whereas you may be a bit slower in other areas and find that being passed here takes longer. Your progress is checked by your supervisor. Meetings with supervisors may be little more than a chat or they may be more formal, again depending on the size of your firm. What is important to understand is that your progress is being continually checked against your plan so that updated plans can be developed.

It is essential that you are fully aware of the qualification you are aiming to achieve and exactly what the training entails. NVQs are divided into units and each unit consists of two or more elements. A major NVQ qualification, such as performing engineering operations, consists of core units, support units and optional units. Additionally, you need to achieve key skills in 6 areas.

NVQ2 - performing engineering operations

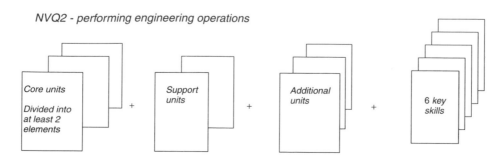

You need to appreciate that becoming qualified depends on you, as you are the one who must show that you have reached the skill level required and that you know and understand what you are doing and can demonstrate your knowledge and understanding by answering questions. Your supervisor or training advisor should ensure that you have been told what is required of you and explained what is needed for you to become qualified.

It is important that you feel that you are able to talk to the people whose job is to ensure that you are being trained properly. There is no disgrace in admitting that you need answers to some questions or that you do not always understand everything you have been taught. Also, do not be afraid to tell people when you are finding work easy or enjoyable.

Complete the following exercise to show that you have a good overview of your training. If there are any questions that you cannot answer, then make sure you get the answer from your supervisor.

EXERCISE 2.2   Knowing about your training

Where are the details of your qualification and aims kept? Are they up to date?	
Apart from your supervisor, what are the names of the people who are responsible for your training? (You may include people at college, key skills trainers and your training advisor)	
By what date (approximately) do you expect to complete your NVQ2?	
Explain what you should do when you feel you are ready to be assessed:	
If you are concerned that you do not understand something or need help, what action should you take?	

Identify any engineering activity that you are not allowed to perform under a specified age, e.g. drive a forklift truck:	

Witness testimony

The above exercise has been completed satisfactorily by . . . .

Signed . . . . Job title . . . . Date . . . .

Finally, it is important that you keep copies of your training plans, signed by you and your supervisor. The fact that you are doing this and completing the exercises throughout this book shows that you understand the nature of your qualification and what is required of you.

Dress, presentation and behaviour

Being correctly dressed and well presented can positively influence your performance at work, your safety and how other people see you.

If you are clean, tidily dressed and well presented, you will feel better in yourself and will be able to work more effectively. Being correctly dressed means wearing clothes appropriate for the work. Clothes should be practical, protective, safe and give some indication of your role.

In the engineering workshop, appropriate clothing refers to special protective clothing, usually overalls, eye protection and protective footwear. Under the Health and Safety at Work Act 1974, it is the law that appropriate clothing must be worn in engineering workshops. (See Chapter 1)

Staff employed in other areas are also expected to dress appropriately: e.g. a receptionist will dress smartly, canteen staff will wear white overalls and caps and a director usually wears a suit.

Being well presented is not only about your appearance but also about your manners such as speaking clearly, being positive and showing an interest in what you do.

Depending on the type of job you are doing, you may or may not have to wear protective clothing at work. The section on the Health and Safety at Work Act outlines the different forms of clothing that have been designed to reduce the effects of accidents. However, it must always be remembered that clothing is only of use when it is being worn properly and that it must be to the correct specification. Eyes, ears, fingers and toes are very precious and cannot be replaced if severely damaged. Also, if protection should be worn, but is not being worn and an accident occurs, then the consequences for the employee can be very serious. There is nothing macho about not taking care of yourself. Skilled craftspeople wear the right clothes for the job and are not self-conscious about doing so.

Fill in one of the next two exercises depending on which applies to you:

EXERCISE 2.3a   Appropriate dress

This exercise is for those who regularly need to wear safety clothing.

Describe the appropriate clothing you wear that is required by your workplace:	
Who is responsible for ensuring that you are wearing the correct clothing?	
Who is responsible for ensuring that you are provided with the correct protective clothing?	
What action should you take if any item of clothing is damaged or needs replacing?	

Witness testimony

The above exercise has been completed satisfactorily by . . . .

Signed . . . . Job title . . . . Date . . . .

EXERCISE 2.3b

This exercise is for those who have to wear safety clothing occasionally.

You see a sign with a safety helmet on it as you enter part of a building, a workshop or a building site. What must you do if you have to enter this area?	
If you are in a stores area and you see a truck reversing, what will you hear?	
If you are in an area where safety clothing is required, who is responsible for making sure that you wear the right clothing?	

Who is responsible for ensuring that protective clothing is provided?	
What action might your employer take if you were found in a 'hard hat' area and not wearing a hard hat?	

Witness testimony

The above exercise has been completed satisfactorily by . . . .

Signed . . . . Job title . . . . Date . . . .

Time keeping

Are we waiting for Alan again. I'm fed up with him!

'Time is money', is what we are often told and 'a fair day's work for a fair day's wage' is another saying we may hear. These sayings are not out of date. When someone is part of a team and the team needs everyone present to start work, there is nothing more annoying than always having to wait for the same person.

An employer could argue that if someone is frequently late, the effect is the same as stealing because the person is being paid to be present. Another issue in this category is 'time-off' work. If someone is genuinely ill then they should not come to work because they could infect other employees or they may feel so ill that they are distracted and are a danger to themselves and to others. Only you know how genuinely ill you feel, but be aware that frequent absences are noticed. Nowadays, employers usually enquire about the number of days taken off in the previous year when asking for references, and they also ask about punctuality and reliability.

EXERCISE 2.4

Statement by supervisor

This confirms that you arrive at work punctually and can account for any days taken off.

Supervisor's signature. . . . .   Date . . . .

Tools and equipment

Part of being positive is **being responsible**. Just as we may say 'a bad workman blames his tools', it is also true that a good workman knows the importance of keeping tools and equipment in good working condition. Badly maintained equipment can be very dangerous because it is not up to the job it is designed to do. A trainee needs to be shown what faults to look for with regard to the equipment being used and how to carry out basic maintenance. It is not a trainee's job to do more than this unless told otherwise. For example, electrical equipment needs to be checked regularly by a trained electrician to ensure

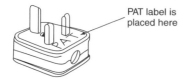

PAT label is placed here

that plugs are wired safely, that earth leads are in place and that the equipment is not hazardous.

You may find stickers with a date and signature attached to plugs on portable appliances. This shows that the appliance has been portable appliance tested (PAT) and these stickers should never be removed.

Similarly, all pressure vessels, e.g. steam boilers have to be regularly tested to make sure that they are safe. These safety checks are required by insurance companies before they issue the required insurance cover that an employer needs. All of these checks are carried out by people who have been trained to do so. A trainee should notice these signs and be aware that if they are concerned about the safety of any equipment then they must report them to the supervisor.

Tools

Tools should always be in good condition and must be taken care of. Large organisations may have regular in-house checks for inspecting and calibrating the accuracy of measuring instruments in a special area called the 'standards room'. Smaller organisations may send their tools away to be calibrated.

Please, what do I do with broken tools?

Organisations usually have a procedure for reporting deficiencies in tools. If this is the case in the place you work and you suspect that the tool you are using is defective, you should follow the set procedure. If there is no set procedure to follow, you should ask your supervisor what to do. What you should not do is just put the tool back and take out another one. This would mean that someone else would have to do your work and you would not find out what was the right thing to do.

Equipment

If you are working on a job and the equipment you are using no longer works as you would expect, do not attempt to repair it. If the equipment is a machine, then switch it off and immediately report the matter to your supervisor, making sure that you give as much information as possible. Watch how the supervisor deals with the problem. The problem may have been that you were using the machine in the wrong way, but the machine may need repair and the correct procedure here is usually to contact the maintenance department.

Materials

Defects in materials may also cause you problems and it is not the purpose of this section to deal with why materials may be defective. However, you do need to know that you cannot do a good job if the material you are working on is defective and that you need to find out the correct procedure for discarding material and getting material that is suited to the task.

EXERCISE 2.5   Reporting deficiencies in tools, equipment and materials

Trainee's evidence

Describe the set procedure for reporting defective tools, equipment and materials.

Briefly describe an occasion when you correctly reported a suspected deficiency in a measuring tool, some marking-out equipment or a material.

Supervisor's confirmation

Confirmation that . . . . is familiar with and reports all suspected deficiencies in tools, equipment and materials according to the set procedures.

Signed by supervisor. . . . .   Date . . . .

Working with other people

Respecting the views of others

Being at work usually means working with other people. You can choose your friends but not your colleagues (or relatives!). The 'other people' can make being at work very pleasant but on occasions these people can also cause problems. Trainees can expect a certain amount of teasing, but teasing that gets out of hand becomes bullying. Probably the best advice here is to treat people the way you would like to be treated. Older employees expect to be respected and many will not tolerate 'lip' or 'cheek'. Make sure you know how people should be addressed. You may find that first names are fine in some circumstances but that some people prefer to be addressed more formally.

At work, there are usually set procedures but sometimes part of a job may be open to interpretation. If you disagree with a colleague's interpretation and method, there is no point in jumping in and arguing about it. Listen to their views and consider carefully and ask your supervisor for advice if there is still uncertainty.

I put your tools away, Mr. Brown. Is there anything else you want me to do?

Remember that those in charge have been appointed to their position on merit and are usually experienced. Pay attention to your supervisor and other personnel in senior positions and follow their advice.

Be prepared to be helpful, and if you want help, make sure that the person you ask has time to help you and that you ask in a polite way. When the help has been given, make sure that you say 'thank you'. Also, think how you could help yourself; show some initiative but be careful that you are not doing anything that you should not do. It may seem easier for you to ask for help, but when someone else is trying to get their work done, it can be quite annoying to have a trainee constantly asking questions – therefore, think about helping yourself. There are lots of ways of doing this and they are covered in Unit 3, which is about using and communicating technical information.

Respecting others' property

There is nothing worse than not being able to trust the people you work with. If you need to borrow tools or equipment from a workmate, it is important to ask properly, to look after them appropriately and return them promptly in the some condition as you received them. This contributes to good team work. If you do not work in this manner, you may lose the trust of the people you work with, who will be much less likely to lend you things in the future.

The tools used from the company or college store should also be looked after properly. Most storekeepers tend to treat these tools as their own and woe betide anyone who does not look after them properly. Also, if you always return items promptly and in good condition, you will make a friend of the storekeeper who will ensure that tools are available when you need them.

EXERCISE 2.6   Respecting the property and views of others

The following exercise can be completed when you feel you have worked with people long enough and have borrowed tools or equipment and returned them.

Trainee's evidence

Describe a situation in which you regularly borrow and return tools and equipment correctly.

Supervisor's confirmation

This confirms that. . . . . respects the views and property of other people.

Signed by supervisor. . . . . Date . . . .

Teamwork, cooperation and integration

At work, it may be that much of your time will be spent working as an individual under supervision or working constantly alongside colleagues. Whatever the situation, you, your colleagues and your supervisor are members of a working group or team.

Members of any team need to work well together to achieve the same aims. The aim of a football team should be to score goals by skilful interactive play. At work, the aims of the engineering team members should be to work together for the success of the organisation and the well-being of the employees.

Two important aspects of working well in a team or group are cooperation and integration. To be successful you need to:

- cooperate with your supervisor when instructions are issued
- cooperate with colleagues when help and assistance are needed
- cooperate with other individual team members or other teams of workers in the same organisation.

That should tell them what they need to know

For example, if you are working in an engineering drawing office, you should show cooperation not only to those around you in the drawing office but also to the craftspeople who will have to work from your drawings. Therefore, on a particularly difficult component, you should help those who have to interpret the drawings by adding extra notes or details to the drawings.

A further example of cooperation occurs when people are working in shifts. You may operate a machine for 8 hours but someone else is going to work at the same workstation as you for the next shift. You will not be very happy if you come to work with a target to meet and find the machine is not set correctly or material that should be at hand has all been used up and you need to collect more. Basically, in

such a situation, leave things as you would wish to find them. If there is ever a complaint, you can be sure it will not be against you and, in addition, you may feel in a position to ask for more cooperation from your shift colleagues.

To be cooperative with the staff outside your immediate work area, it is necessary to have some understanding of their work. Knowing what they do will make cooperating with them easier and will also give you a fuller insight into all the activities of your organisation.

Many companies ensure that trainees achieve a good understanding of all aspects of the company's work by moving the trainees around from one department to another throughout their training. This way of working will certainly help you to cooperate in all aspects of work and it will also help you to explain about the firm to newer staff and visitors.

In the past, people were often asked to carry out repetitive tasks every day. This way of working often led to boredom and low output. The modern approach is to work in teams in which people share the jobs and are required to do the repetitive tasks only when it is their turn. The team approach also encourages those with leadership qualities to encourage everyone to work together. Working in a team also means that you need to trust the other team members.

Teamwork can have its drawbacks. Most people have strengths, things that they are good at, and they also have weaknesses. Getting the best out of a team means making the best use of people's strengths and minimising their weaknesses. It also calls for tact and understanding, for example, you might be frequently asked by someone to lift things when you are working with them. Before you complain too bitterly or make a fuss, just check that this person does not a have bad back!

EXERCISE 2.7   Showing that you can work with other people

Trainee's evidence

Describe briefly how you have to work with other people in a team.

Describe briefly how you have talked and cooperated with other people outside your team.

Describe briefly how a disagreement has occurred and how it was resolved constructively.

Supervisor's confirmation

This confirms that . . . . regularly cooperates with colleagues and visitors.

Supervisor's signature. . . . .   Date . . . .

Working in an organisation

Below you will see a diagram showing how a typical company may be structured. Try and identify your place in the company. It may also be fun to think where you would like to be in ten years' time! Looking at a diagram such as this should help you to get a better idea of who does what and the total number of people you need to learn to work with.

EXERCISE 2.8   Organisation chart

Draw a company organisation chart like the one below for your workplace. Where possible identify people by name and show where you fit into the scheme of things.

If the organisation for whom you work is very large, then show part of it. If the organisation is very small, then give as much detail as you can and try to write notes about the various jobs that people have to do.

Typical engineering company management structure

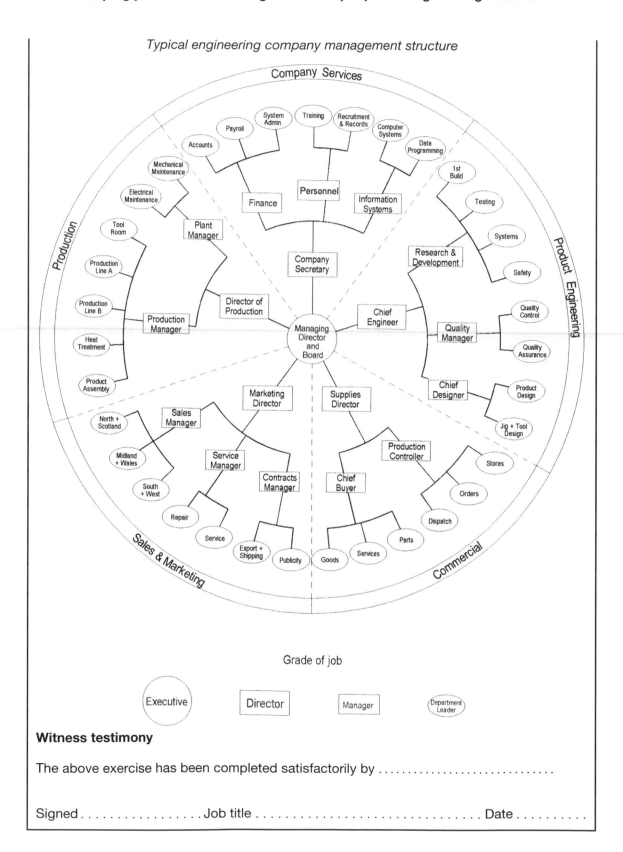

Grade of job

Executive   Director   Manager   Department Leader

Witness testimony

The above exercise has been completed satisfactorily by ....

Signed . . . . Job title . . . . Date . . . .

Reporting difficulties in working relationships

The majority of people realise the importance of good working relationships and will try to get on with others as best as they can. However, in working environments, even though there are good working relationships, problems among people sometimes arise.

Problem	Example cause
Inconsideration	Ignoring other people's feelings and needs.
Clash of personalities	Difficulties or differences related to jealousy.
Unfair treatment	Treating people in an unfair or unacceptable manner.
Lack of trust	Arising from a lack of respect for people and property.
Rejection	Deliberately or inadvertently leaving people out.
Harassment	Picking on someone, bullying and continual sarcasm.

I don't like what's going on.

In most cases, the problems are minor and, with a lot of give and take, they can usually be resolved by discussion among the people involved. Colleagues who may have noticed a problem – possibly someone being bullied – will often step in to help sort things out.

The usual way of sorting out things is by talking about the issues with the people involved. It is also worth noting that people who are respected in a workplace are not people who treat others badly. There may be peer pressure to 'go with the crowd' as far as someone else is concerned, but do you really want to make someone else miserable?

It is worth trying to sort out problems there and then, by discussion; but discussion does not involve shouting. However, if this approach has failed or is not practicable, it is important that difficulties among colleagues are sorted out and this may mean reporting the difficulty.

In many organisations, there are set procedures for reporting problems; the procedures usually involve some or all of the following persons:

- supervisor
- safety representative
- personnel officer – human relations manager
- department manager
- organisation counsellor
- organisation mediator.

All these people have an interest in maintaining good working relations, both at work and elsewhere. They are there to help you because if you are having problems, you are probably not working as well as you are capable of, and this could affect other people.

In addition to Company policies, there are Employment laws. These laws ensure that people at work are treated fairly and not exploited. In 1970, the Equal Pay Act was brought in to ensure that men and women doing the same job must receive the same pay. The Sex

Discrimination Act 1975 and the Race Relations Act 1976 make it an offence to discriminate between individuals on the grounds of gender or race. The Equal Opportunities Commission looks into breaches of employment law where people think they are being discriminated against, for example, people in wheel chairs now have rights to demand access to workplaces and they cannot be discriminated against because of their disability. Cases of discrimination do not go to the police; they are dealt with by civil law.

Taking part in discussions

When things have gone wrong, people discuss the events. We all take part in discussions – often we just call them conversations, and it is very much a part of everyday life. In the work situation, we may be called upon to add our point of view, our experience or our knowledge. Discussions are often held to make matters clearer, to resolve differences or to improve the situation. In engineering, you may be asked to comment on the operation of a new piece of equipment or a new process. A whole process may be reviewed to make it safer or more efficient, and as part of the team, your views will be required as well, so you need to take part in the discussion.

Discussions often have different stages but the outcome is often that after everyone has had an opportunity to have their say, a decision needs to be taken. During the discussion phase, people need to listen to what is being said as well as make contributions. We all know there are people who monopolise the discussion and will not listen to other people's point of view; this is not necessarily going to produce the best outcome.

To be a useful contributor to a discussion, you need to listen to what other people have to say, note the main points they are making and then be prepared to make relevant points of your own, which take the discussion forward. Sometimes you need to take clues from a person's behaviour, facial expression and body language. Some people are not as assertive as others, yet they may want to make a point; by being aware of their nonverbal communication, you can show that you are aware of their needs as well as your own.

Discussions at work are not about you getting your own way but about making the right decision, which will not be arrived at as a result of a shouting match. An employer will be far more impressed with someone who listens, adds useful points, summarises, makes points clear and who genuinely contributes.

EXERCISE 2.9   Taking part in a discussion

The following exercise is designed to encourage you to discuss possible problems that may arise in work situations. To obtain key skills in communications **you must contribute to a discussion about a straightforward subject.**

Topic(s) under discussion

Date(s)

What you need to know and do	Assessor's initials	What the evidence consists of. (Possible evidence suggestions in italics.) Assessor can highlight or add additional comments
Make clear and relevant contributions in a way that suits your purpose and situation by choosing appropriate and varied vocabulary and expressions to suit your purpose		*Sensible comments made.* *Relevant comments made.* *Explanations are offered if technical terms are used.*
You should adapt your contributions to suit different situations		*Full explanation given if needed and if candidate is obviously an 'expert'.* *Tone of voice is suitable, not too aggressive or bullying. Swear words should not be used.* *Points made without shouting.* *Further explanation requested.*
Listen and respond appropriately to what others say by showing that you are listening closely and responding appropriately		*Taking an obvious interest in what is being said by listening to a point and then making a relevant comment.* *Showing an awareness that others are taking part in the discussion as well.* *Picking up on points made by someone else.* *Allowing someone else time to make a point without butting in.* *Not shouting down another person.* *Actually asking someone who is not an active participant in the discussion what they think.*
Help to move the discussion forward by identifying the speaker's intention and by generally moving the discussion forward		*Making comments like 'where do we go from here?' or 'what's the next step or thing to do?'* *Helping to make clear what another speaker has said.* *Establishing that a point has been made.* *Summarising where the discussion has got to and what still needs to be resolved.* *Keeping voice pleasant and not shouting ensuring that everyone has a chance to make a point.*

Witness testimony

This confirms that ....took part in discussions and that key skills requirements as listed above have been met.

Signed................Job title ....Date.........

Look through the following case studies and discuss with other people in your group how you feel they should be sorted out.

Case 1 You work with someone who is regularly late for work. This affects you as it means that you are the one who has to do most of the preparation. The same person is also quite happy to 'skive' off early and leave you to do the clearing up. The excuse given is that this is what happened to him when he was in training.

The names of the people in our group were:

We feel the best way to sort this out is to:

Case 2 You work as part of a team and it is near the end of the shift. You have finished your work and were hoping to get 5 minutes off early but your supervisor asks you to help someone else. This person was late for work today. How are you going to cope with this situation?

The people who discussed this were:

What we feel we should do is:

Case 3   You notice one of your more experienced colleagues taking short cuts and using a machine without the proper guards in place. You also know that this machine is left in this condition for the next person to come on shift to use. What should you do?

The people who discussed this were:

What we think you should do is:

Case 4   You notice someone of an ethnic origin different from most of the workforce being given quite a lot of 'grief' to the extent that s/he is thinking of leaving. What should you do?

The people who discussed this were:

What we think we should do about this is:

Case 5   One of the trainees, who is already quite skilled and doing NVQ Level 3, is deaf. He lip-reads very well but obviously is unaware of remarks made. You notice that some people comment quite openly about him, knowing that he cannot hear them, and some of the things said are not very pleasant. He applies for further training but is turned down basically because he is deaf. Are there laws to prevent this type of discrimination?

The people who discussed this were:

Our views on this are as follows and we think this situation is covered by the law in this manner:

Being helpful

When we start a new job, we hope that there will be plenty of people around who will tell us what to do, where to get equipment and generally make life easier for us. What we also need to remember is that a time will come when we are in a position to provide the same help to other people.

One way in which you can be helpful is by making sure you know how your organisation operates. By doing this, you will make life easier for your supervisor because you will be thought of as someone who can be relied upon. All organisations need to communicate with the people in the organisation and there are various ways in which this can be done; the following may be used in your workplace:

- Verbal instructions and information given out by supervisors or other line managers.
- Written instructions. These may be a variety of forms and could include requests for information from the Human Resource Manager (Personnel), right through to Process Information, which is the written sequence of operations and notes for production.
- Bulletins and newspapers. Some large firms have their own in-house newspaper or bulletins, which may be pinned to notice boards.
- E-mail or electronic mail is becoming increasingly common but suffers from the same problems as other forms of communication in that there needs to be a check to ensure that everyone has read the information.

Effective communication

To be effective, communication needs to be a two-way process. All of the ways of communicating outlined above fall into the category of one-way information and hence have limited use. In engineering, it is essential that instructions are not only issued but are also understood. Most children would have played the 'whispering game' in which one person whispers something to another and the whispering goes on to the end of the line where the last person repeats what was heard. What this person says usually causes everyone to laugh, as the message has been corrupted along its path. Effective communication ensures that the person receiving the message receives the correct message and is able to ask questions, query information or generally feel free to discuss what has been received.

Look at the following exercises – Exercise 2.10 asks you to identify the types of communication methods used in your workplace, while Exercise 2.11 asks you to think about how to ensure that communication is effective.

EXERCISE 2.10   Identifying the types of communication methods present at work

Communication method	Used in your workplace? Yes or No	Notes (name examples or give other information)
Notices	Yes	Pinned up in canteen. E.g. health and safety information.
Verbal instructions		
Written instructions		
Written requests for information		
Firm newspaper		
Electronic mail		
Bulletin board/Notice board		

Witness testimony

The above exercise has been completed satisfactorily by . . . .

Signed . . . . Job title . . . . Date . . . .

EXERCISE 2.11   Thinking about effective communication

Communication method	One-way or two-way	Used by you in the following way:
Discussion with supervisor	Two-way	To discuss progress, to clarify points I do not understand.

Company newspaper		
Written instructions		
Request to senior colleague for help		
Request to tool room for equipment		
Letter of application for a job		
Telephone call to a company for technical assistance, e.g. COSHH information		

Witness testimony

The above exercise has been completed satisfactorily by . . . ., who always seeks assistance in a courteous and polite manner.

Signed . . . . .Job title . . . . Date . . . .

You as a communicator

It is important to remember that you also need to provide information; you need to confirm to your supervisor that you have understood what you are being told. As part of being a positive member of a team, you need to respond quickly to requests for information from you. You may also need to offer to take the initiative and tell other people what is going on. Also, you should be aware that you need to inform people if something you are about to do could cause them inconvenience, for example you may have to shut down a shared electricity supply while you make connections. Obviously, you must tell others when this is about to happen as apart from being inconvenienced they could also be in the middle of a delicate operation that could be ruined if the power went down. Similarly, if you are about to do anything else that could alarm people or cause any amount of danger, then it is your duty to inform others of your intended operations.

Sometimes we also have to fill out official forms, such as an accident or incident form, which are very important as they are for

Mr. Ali, I've got those drawings in front of me.

official purposes. An accident form effectively notifies management that an accident or incident has occurred. All employers have to have insurance that covers everyone who works on or visits the site. So filling in an accident form is important for insurance purposes and it is in everyone's interest that these are filled out when the accident happens or as soon as possible afterwards. They need to be accurately filled in with as much detail as is requested.

EXERCISE 2.12 Showing that you can communicate effectively with those who are helping to train you

Give an example of when you were worried about some aspect of your training and you sought advice from your trainer or supervisor:	
Explain briefly why you should confirm instructions you are unsure of:	
Explain why using approach (a) shows that you are a better communicator than when using approach (b): (a) 'Excuse me, could you explain that again please?' (b) 'Oi you, how do I do this?'	
Give an example of when you took advice from someone in order to improve your work:	

Witness testimony

The above exercise has been completed satisfactorily by . . . .

Signed . . . . Job title . . . . Date . . . .

EXERCISE 2.13   Accident reporting

Read through the following report of an accident and then complete either the accident report form in this book or one provided by your supervisor. You will notice that another form is mentioned as well, which is a risk assessment form. Risk assessments are carried out to identify potential hazards and the risks associated with carrying out specific activities. See the Safety section.

Accident Report

The following is the report of an accident that occurred on a building site after a severe storm:

Name of Employee: Stu Pid

Address: 4 Elm Street, Washburton, Cheshire

Date of Birth: 13 May, 1964

Statement from Mr. Pid:

When I got to the site in the morning after the storm I noticed that quite a lot of damage had been done to the top of the brickwork I had been working on the day before. So, first of all I filled a barrel with bricks and then I went up to the work and rigged up a block and tackle. Then I pulled up the barrel to carry out the repair work. There were some bricks over so I lowered these to the ground. Unfortunately the rope slipped and the barrel went down too fast and burst when it hit the ground. As I was acting as the counterweight at the top I was now heavier than the broken barrel and fell from the top onto the pile of bricks at the bottom. On my way down I was hit severely on the shins by the broken empty barrel on its way up. When I hit the bricks at the bottom I broke my ankle. The pain was so bad that I forgot what I was doing and let go of the rope. The next thing I knew was that the empty barrel landed on my head and knocked me unconscious.

Statement from the Investigating Health and Safety Officer:

I was called to the site at Grangemouth at 8 a.m. on the morning of Friday 13th February 2000 by the site manager. A passer-by had witnessed the accident and luckily carried a mobile phone so had called for an ambulance. Mr Pid was treated by paramedics at the site before being taken to Grangemeouth General Hospital for treatment for a broken right ankle, concussion and cuts to his left leg and face. I later visited Mr Pid in hospital. Mr Pid told me that he had been worried about storm damage and being quite concerned about the unstable state of the brickwork had decided to start work on his own rather than wait for his work mate. I have completed the Accident report form and have also carried out a Risk Assessment, although procedures are in place for normal working, as the storm represents a situation we have not considered before.

Action 1
Use information from the above report (which was given to the site Health and Safety Officer by the injured man in hospital) to fill in the accident report form which can be found as Appendix XV at the end of the book.

Action 2
Imagine you are the site H & S Officer. Carry out a risk assessment using the form provided at the end of the book in Appendix XIV and make recommendations as to the procedure to be followed.

Witness testimony

The above exercise has been completed satisfactorily by . . . .

Signed . . . . Job title . . . . Date . . . .

Another form of communication that is very powerful is 'nonverbal communication'. This form of communication includes:

- facial and body gestures
- your appearance
- the way you position yourself in relation to someone else, e.g. turning your back on someone
- giving someone your attention or obviously not listening
- making eye contact.

Very few of us probably appreciate just how much we communicate in this nonverbal way, but the messages we give out are very obvious.

EXERCISE 2.14   Look at the two pictures below and note four points each for good and bad communication

(a)   (b)

(Include both verbal and non-verbal communication.)

Good	Bad
1.	1.
2.	2.
3.	3.
4.	4.

Witness testimony

The above exercise has been completed satisfactorily by . . . .

Signed . . . . Job title . . . . Date . . . .

Talking to Visitors

Outsiders visit organisations all the time. For security purposes, visitors are normally required to sign-in at the reception and often have a security tag with their name attached, but this is not always the case. The visitor may be a salesperson, a shareholder, a potential employee, your training advisor or even royalty! (We usually know when VIPs are expected.) You never know when you will be required to talk to a visitor, if only to give directions.

It is important that you appreciate that whenever you talk to a visitor you are representing your organisation and you should speak politely. On occasions, there will be official visits and you may have to explain what you are doing, how you are being trained and you may even be asked to demonstrate how you carry out your work activity. Your firm will expect you to do this in as positive a way as possible. So:

- speak up and speak clearly
- be polite
- insist on safety precautions being taken if necessary, e.g. ask the visitor to put on safety glasses
- always be positive – if you do not know the answer to a question explain that you cannot provide the answer, but point the visitor toward your supervisor for a response
- always present your organisation in a good light; remember that it is bad manners to embarrass a visitor with private grievances.

EXERCISE 2.15 Providing evidence of good communication to visitors

Trainee's evidence In the space below outline an occasion when you were called upon to talk to a visitor to your workplace.

Supervisor's confirmation

This confirms that . . . . can be trusted to talk sensibly and positively to visitors to the workplace.

Supervisor's signature. . . . .   Date . . . .

EXERCISE 2.16

(Application of Number – Assignment 2. This exercise is specifically included to enable requirements for key skills to be met.)

Jay is 19 years old, passed the driving test at the first attempt and has driven the family car for a year without any convictions or accidents. Now Jay wants to buy a car and has taken into consideration:

- money saved to buy the car
- make of car
- size of engine–for road tax purposes
- fuel consumption.

Task 1

Look carefully at the following data and answer the questions that follow:

Make of car	Engine size	Road tax below 1.1 l £100 above 1.1 l £155	Insurance (3rd party only)	Fuel cost of driving 6000 miles
Fiesta	1.2 l	£155	£870	£550
Escort XR2	1.6 l	£155	£1700	£626
Skoda Fabia	1.0 l	£100	£700	£504
Golf GTI	1.6 l	£155	£2500	£600

(Insurance information supplied by Wyedean Insurance, Monmouth)

(a) Which car will cost Jay the most in insurance?

(b) What size of engine costs the least in road tax?

(c) Which car gives the best fuel consumption?

Task 2

Jay has a maximum of £2000 to spend and finds a 1.2 l seven year-old Fiesta at £700. Complete the following:

(a) The insurance will be: third party   £

(b) The road tax will be: £155 for 12 months   £155

(c) The car will cost   £

(d) The MOT will cost   £32.50

(e) It needs 2 new tyres @ £18 each   £ _____

(f) The total amount Jay will spend will be: £

(g) Draw a bar chart to show: (i) the total amount of money Jay will spend and (ii) how the total cost is broken down into the categories given above (using one bar for each cost).

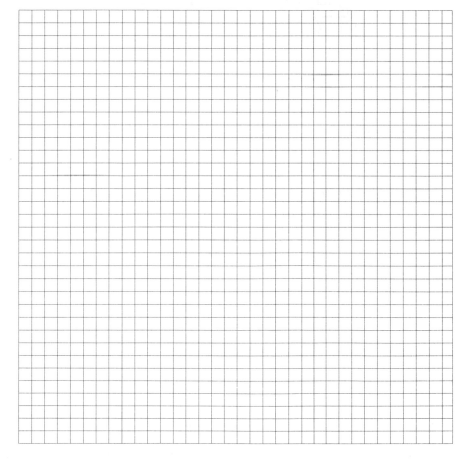

Task 3

Jay can expect to spend £550 in petrol for driving 6000 miles in one year. Approximately (to the nearest £) how much will Jay spend on petrol per month?

Jay is paid weekly. How much will petrol cost per week (again to the nearest £)?

Task 4

Look at the information below that refers to the Ford Cougar. The 2.0 l has a 16-valve engine while the 2.5 l has a 24-valve V6 engine. Part of the performance of the two engines is shown below.

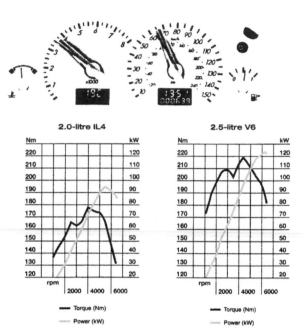

(a) What is being measured on the x axis?

(b) At what rev/min is the maximum torque produced for each engine?

2.0 l

2.5 l

(c) How much extra torque does the V6 produce at 2500 rev/min?

(d) What is the major difference between the V6's power (kW) graph shape and V6's torque (Nm) graph shape?

Task 5

Look at the following information about new and used car prices.

(a) On a one-line graph, plot the age of a car from new (0 years) to 5 years. Make sure you label each line of the graph with the car it represents. Also make sure that your graph has a title and that you have labelled the axes.

Car	New N reg. (Prices rounded off)	I year	5 years Fair price
BMW Coupe	£19 000	£13 950	£8100
Renault 2l Exec.	£18 200	£9560	£4100
Rover saloon	£17 900	£7400	£2200
Subaru 2000	£18 000	£17 000	£3900
Vauxhall Auto	£17 800	£9700	£4000

Cars and prices have been chosen to provide a sample that is comparable.

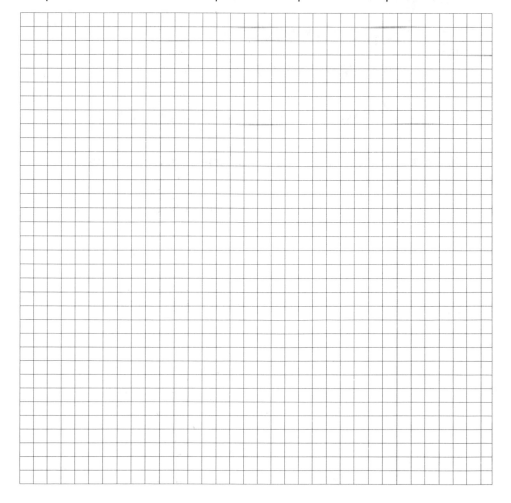

(b) Which car has held its price best?

(c) For the Vauxhall:

 (i) What is the difference in price between a new and a five year old one?
 (ii) Approximately by what percentage does its value drop in 5 years?
 (iii) By what percentage has the price fallen between being new and 1 year old?

Task 6
Jay had to buy two new tyres for the Fiesta before it went for its MOT test. Good tyres are essential for safe braking. Look at the following information regarding safe stopping distances and for each speed calculate the ratio of thinking time to complete stopping distance: (Do your calculations to the nearest whole number)

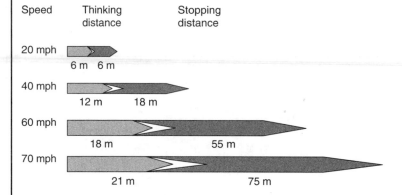

Speed	Thinking distance	Stopping distance
20 mph	6 m	6 m
40 mph	12 m	18 m
60 mph	18 m	55 m
70 mph	21 m	75 m

(a) At 20 mph, the ratio of thinking distance to complete stopping distance is

(b) At 40 mph, the ratio of thinking distance to complete stopping distance is

(c) At 60 mph, the ratio of thinking distance to complete stopping distance is

(d) At 70 mph, the ratio of thinking distance to complete stopping distance is

Witness testimony

The above exercise has been completed satisfactorily by ....

Signed . . . . Job title . . . . Date . . . .

3 Communicating using engineering drawings

Exercise checklist

(Ask your assessor to initial the lower box for each exercise when you have completed it.)

Exercise no.	3.1	3.2	3.3	3.4	3.5	3.6	3.7	3.8	3.9	3.10
Initials										
Date										

The most effective way of communicating the size, shape and details of a single component or an assembly is to produce an accurate drawing of the required item. Draughtsmen produce drawings, and after being checked and approved they are copied and distributed. They are used for the costing, planning, manufacture, maintenance, marketing and sales. All the features of a component should be clearly shown on a drawing so that the drawing can be 'read' by any engineer worldwide.

There are a number of guidelines for engineering drawing layout, forms of projection, types of line, dimensioning methods and symbols, which should be observed. These guidelines are published in British Standard documents, the most important being BS 308 'Engineering Drawing Practice'. It is important that these guidelines are followed, so that the information on the drawing is interpreted correctly by everyone who acts on it.

Sketches are used more frequently while discussing ideas and designing ways around problems. A sketch is a quick method of communicating information and uses the same layouts and symbols as are used in engineering drawings, but sketches are drawn freehand and are not usually distributed as widely.

In July 2000, a Concorde crashed during take-off after catching fire because of debris on the runway. 113 people lost their lives. The sketch below was produced to illustrate how the aircraft can be modified to enable its fuel tanks to withstand perforations caused in this way, reducing the risk of fire, by inserting Kevlar panels inside the wings.

Where Kevlar panels will be installed in Concorde's fuel tanks

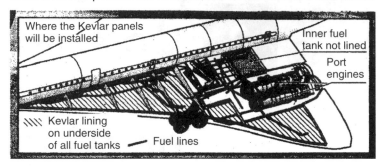

Types of drawing

Sketches are informal and are not necessarily accurate; formal drawings are produced when a design has been finalised. A range of drawing layouts are available to present different types of information in the most descriptive way. First and third angle projection are two of the most common ways of laying out engineering drawings, and are described in detail later. The layout style best suited to the actual component is selected and the drawing is prepared on special paper.

Paper sizes

Paper sizes

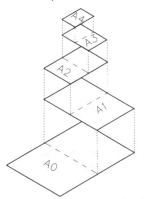

The size of the paper on which all engineering drawings are produced depends on the size and scale of the component being represented. The sheet sizes of metric paper are as follows:

A0	841 mm × 1189 mm (approx. 1 m^2)
A1	594 mm × 841 mm
A2	420 mm × 594 mm
A3	297 mm × 420 mm
A4	210 mm × 297 mm

Note. Each sheet size is half the previous sheet's size.

The paper on which engineering drawings are produced is normally surrounded by a bold border called a frame. At the bottom right-hand corner of the frame there should be a title block that contains technical information about the component. BS 308 requires the title block to contain the following minimum information:

- draughtsman's name or initials
- date of drawing
- projection symbol
- scale ratio
- component title
- drawing number.

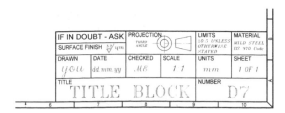

Further information is nearly always needed; for example, it is unusual for a drawing not to specify the material used or the surface finish required. The title block illustrated above contains typical information for engineering component drawings.

Types of line on drawings

Various features on engineering drawings are represented by particular types of line. The table below shows the correct type of line for a number of applications as recommended in BS 308.

Line	Description	Application
▬▬▬▬	Continuous bold	(a) Visible outlines (b) Visible edges
——————	Continuous fine	(c) Dimension lines (d) Leader lines (e) Projection lines (f) Hatching – Adjacent part outlines
∼∼∼∼	Continuous fine irregular	(g) Limit of partial view
- - - - - - - - -	Fine short dashes	(h) Hidden edges – hidden outlines
— - — - —	Fine chain	(i) Centre lines
▬ - — - ▬	Fine chain, bold at ends and changes of direction	(j) Cutting planes

Sample types of lines used in an engineering drawing

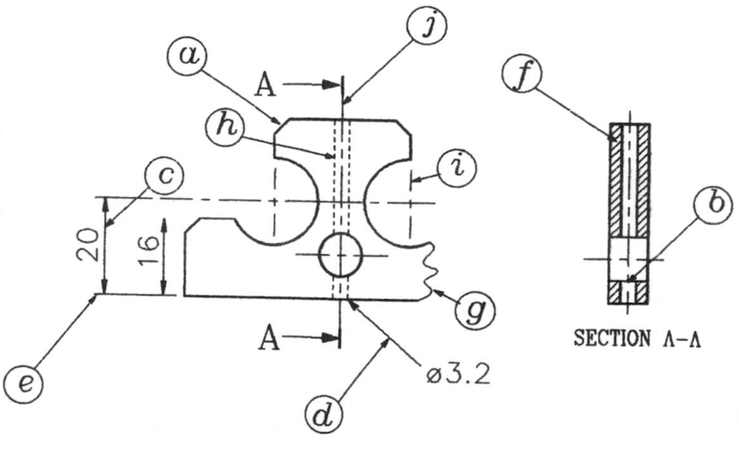

SECTION A-A

Orthographic projection

Designers need to represent components clearly on paper in the form of drawings if they are to be manufactured as required. The preferred method of drawing components for manufacture is to set out two or more views of them on paper in a logical manner using a drawing system called *Orthographic projection*. There are two types of orthographic projection – one is known as *First angle projection*, and the other is known as *Third angle projection*.

First angle projection

First angle projection (sometimes referred to as English projection) is constructed by looking at the component and selecting the faces that reveal the most features. Consider the corner plate shown here. It could be viewed in any of the six planes shown.

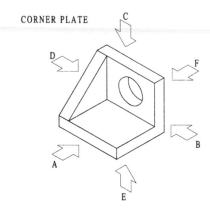

CORNER PLATE

VIEWS

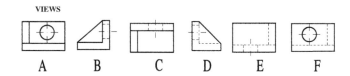

A   B   C   D   E   F

After examining these views, it can be seen that views A, B and C reveal the most visible edges, while views D, E and F have some of the corner plate's details obscured. Generally, three views are chosen and the preferred views A, B and C have been selected here as the most appropriate for a first angle projection drawing of the corner plate.

For the correct layout of the corner plate in **first angle projection**, the procedure is as follows:

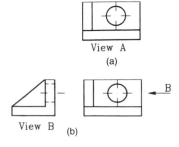

View A
(a)

View B (b)

(a) The elevation considered to reveal the most features is drawn. For the purpose of this exercise, View A of the corner plate is used.

(b) View B is then drawn to the **left** of View A. Note that View B is drawn on the **opposite** side from where it is viewed (arrow B).

(c) View C is drawn directly **below** View A. Note that View C is drawn on the **opposite** side from where it is viewed (arrow C).

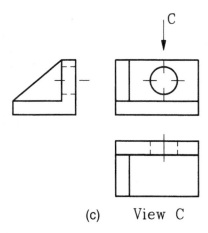

(c)   View C

If View B was considered to be the face with the most detail, then the first angle projection drawing would be drawn as shown:

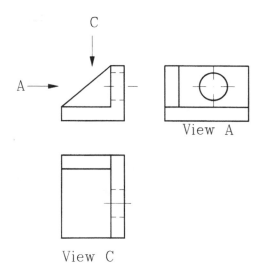

View A

View C

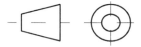

The British Standard symbol for first angle projection is shown here. This symbol is included in the title block whenever first angle projection is used.

Third angle projection

Third angle projection (sometimes referred to as American projection) is an alternative method of laying out the individual views on the drawing page. When using third angle projection, we use the same planes of projection as previously described for first angle projection, but the views are now laid out on the same side as they are viewed from. A drawing of the corner plate in third angle projection is constructed as follows:

View A

The elevation considered to reveal the most features is drawn. For the purpose of this exercise, View A of the corner plate is used.

View B is now drawn on the **right** of View A. Note that view B is drawn on the **same** side from where it is viewed (arrow B).

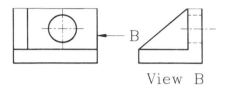

View B

View C is drawn directly **above** View A. Note that View C is drawn on the **same** side from where it is viewed (View C).

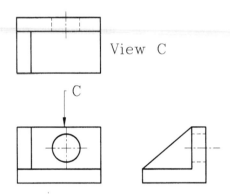

If View B was considered to be the face with the most detail, then the third angle projection drawing of the corner plate would be laid out as shown:

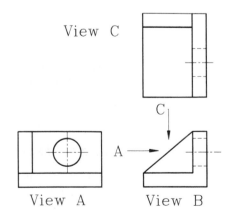

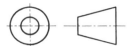

The British Standard symbol for third angle projection is shown here. This symbol is included in the title block whenever third angle projection is used.

EXERCISE 3.1

Task
Examine carefully the Six orthographic drawings of the angle plate shown here. The drawings are shown in either first or third angle projection and some are drawn wrongly. Write in the small box:

1 if you think first angle projection has been used
3 if you think third angle projection has been used
X if you think it is drawn wrongly.

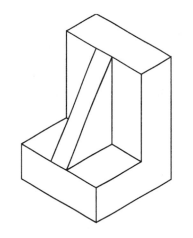

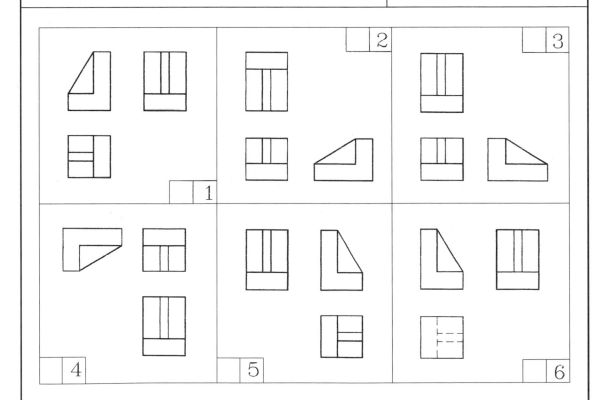

Witness testimony

The above exercise has been completed satisfactorily by . . . .

Signed . . . . Job title . . . . Date . . . .

EXERCISE 3.2

Task

Examine carefully the six orthographic drawings of engineering components below; each has some lines missing. The missing lines may be outlines, visible edges or hidden edges. For your guidance, No. 1 is shown completed in the right margin.

Add to each of the drawings all the missing lines using the appropriate line types.

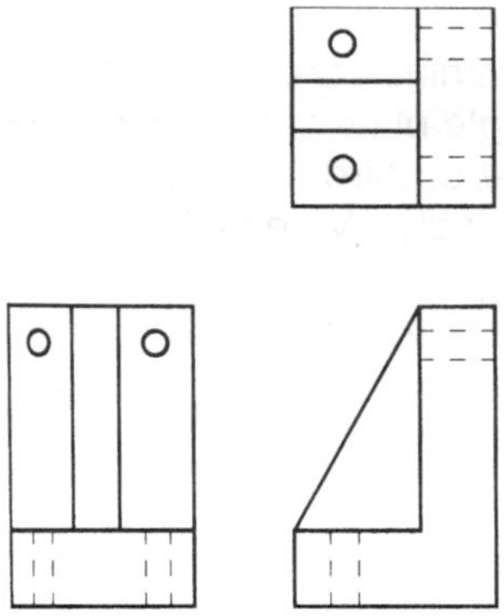

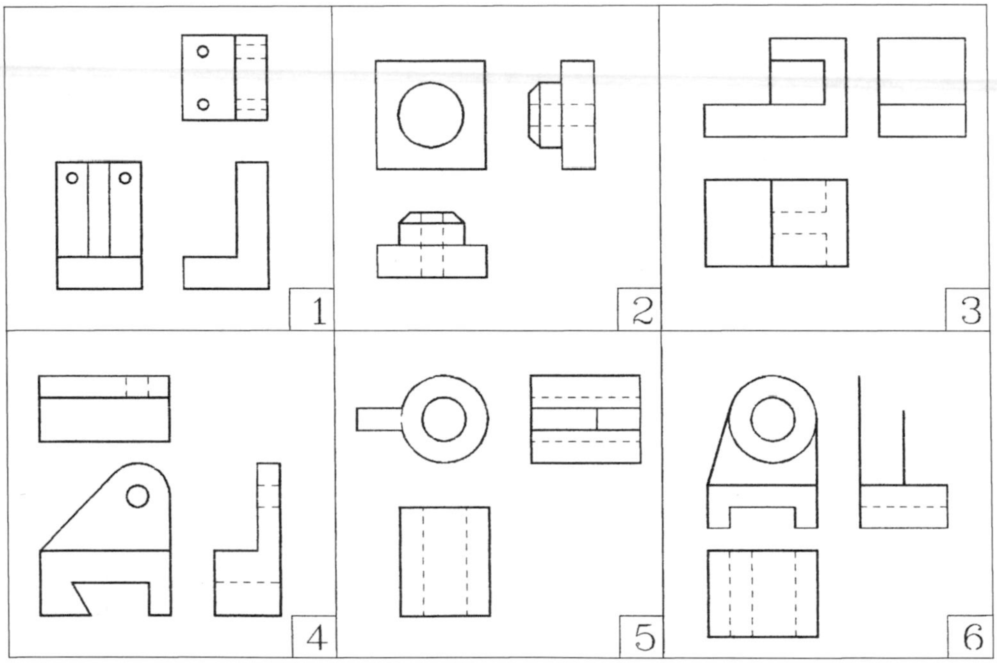

Witness testimony

The above exercise has been completed satisfactorily by . . . .

Signed . . . . Job title . . . . Date . . . .

Dimensioning engineering drawings

An engineering drawing contains information about the shape of a component. It also shows dimensions regarding the sizes and limits to which the component should be manufactured. In this section, all dimensions are in millimetres. Information regarding dimensions must be clear and in accordance with BS 308, the fundamental requirements of which are described below:

- When applying dimensions to a drawing, use thin continuous dimension lines, projection lines and leaders.
- Arrow heads on dimension lines and leaders must be solid and slender.
- Dimension lines should not cross each other.
- When adding notes to a drawing, the letters, numbers and symbols used should be bold and clear and not less than 3 mm high. In general, capital letters should be used.
- When applying dimensions to a drawing, all distances in each direction must be taken from the same face, line or point. This face, line or point is called a **datum**. The height, length and width measurements each have their own separate datum.

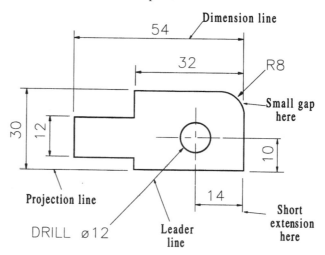

Datum faces
(Also known as datum edges): Most common datum type. All lengths dimensioned from Face A, all heights dimensioned from Face B

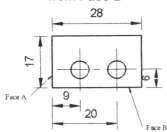

Datum lines
All dimensions are relative to a line. Lengths dimensioned from Line A, heights dimensioned from Line B

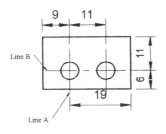

Datum points
The dimensioning of the position of features which all have the same reference point, Point P

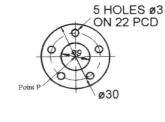

Note. In the three examples above only one type of datum is shown for each drawing. In practice, a drawing may show more than one type of datum, for example, both a datum line and a datum face.

- The symbol Ø preceding a dimension indicates a diameter.
- There are several methods of dimensioning circles. Choose according to the size and location of the circle and according to whether it is a hole or a shaft that is being dimensioned.

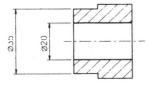

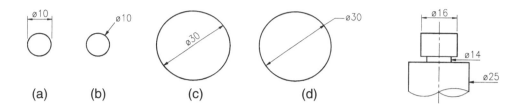

- Dimensions and text should be placed outside the drawing outline wherever possible.

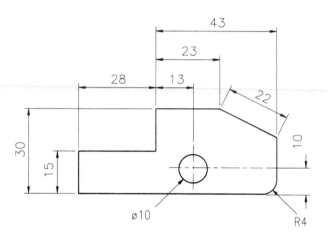

- The text should be written in the middle of and on top of dimension lines so that it can be read from the bottom or the right-hand side of the drawing.
- When dimensioning a radius (curved surface), use a leader and the abbreviation R to precede the size of the curve's radius.
- Each dimension should be given only once and be as close as possible to the feature to which it refers.
- Metric screw threads are specified by the letter M (metric) followed by the diameter of the thread (M8), an × and the pitch of the thread (1.25).
  For example, M8 × 1.25.

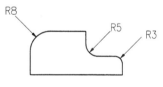

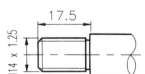

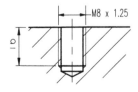

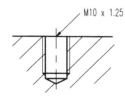

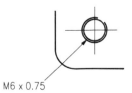

- Holes for fasteners or for locating devices should be dimensioned by one of the alternative methods shown for each type of hole. See figure:

Spot face

DRILL ø8
S'FACE ø18

ø8 S'FACE ø18

Ream

ø8 REAM

REAM ø8

Counter bore

C'BORE
ø12 x 8 DEEP

C'BORE
ø12 x 8 DEEP

Counter sink

90°

ø12

ø6

ø6 CSK AT 90°
TO ø12

- The machining symbol shown below is used to indicate whether a surface is to be machined or not. The symbol is drawn touching the surface (or a projection line from the surface). The machining symbol would normally be about 8 mm high.

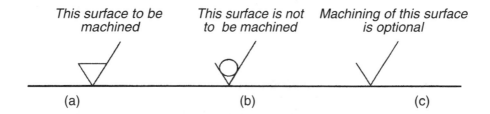

This surface to be machined

(a)

This surface is not to be machined

(b)

Machining of this surface is optional

(c)

This surface is to be machined to give a surface texture of 1.6 μm

- If a specific surface texture is required, the RA (roughness average) value is written on top of the machining symbol as shown. Note that there is more information about surface texture and RA values on page 168 of *Checking completed workpieces for accuracy*. The example surface shown is to be machined to an RA value of 3.2 μm.

Limits of size

There must always be some allowance, however slight, for size on components. This allowance is usually stated in the title block as limits, the general allowable deviation for all dimensions; if special limits are needed for particular dimensions, they must be inserted separately. The correct term for this allowance is the **limits of size**, often called the *limits*. There are always two limits to represent the maximum and the minimum permissible size to which any component feature can be made.

General limits of size are shown on engineering drawings in the title block and they refer to all dimensions. Any feature that has other limits than those indicated in the title block is indicated as shown by one of the methods in the figure.

The BS preferred method, indicating both the limits	Using the ± (plus or minus) symbol	Indicating the limits only
10.5 9.5 (a)	±0.5 10 (b)	10 $^{+0.5}_{-0.5}$ (c)

All the above examples show a feature of diameter 10 mm. The finished component must have a diameter between 9.5 mm and 10.5 mm for it to be acceptable.

The names of the component parts that make up the dimension are:

- the **nominal** size; it is the nearest whole millimetre (or fractional inch) size of the feature; it is 10 mm in the examples above.
- the maximum permissible size; this is called the *high limit* (or top limit); it is 10.5 mm in the examples above.
- the minimum permissible size; this is called the *low limit* (or bottom limit); it is 9.5 mm in the examples above.

The **tolerance**, which is the difference between the limits of size, is 1 mm in the examples above, that is the difference between 9.5 mm and 10.5 mm.

Types of fit

Depending on the actual sizes of mating components, the way they fit together can result in either a **clearance fit**, where there is space between the two mating components or an **interference fit** when the components have to be forced together because the hole is smaller than the shaft.

A clearance fit

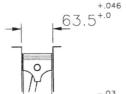

$63.5^{+.046}_{+.0}$

$63.5^{-.03}_{-.06}$

Clearance fit

Whenever components move relative to each other and a clearance fit is required, the hole is designed bigger than the shaft. For example, pistons are smaller than the bore in which they reciprocate. On the drawing the piston must be smaller than the bore.

An interference fit

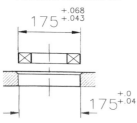

$175^{+.068}_{+.043}$

$175^{+.0}_{+.04}$

Interference fit

When the components are locked together and an interference fit is required, the hole is designed smaller than the shaft. For example, the outer race of ball bearing is an interference fit in its housing. On the drawing, the bearing is larger than the housing.

When drawing mating parts, a designer would refer to BS 4500 ISO Limits & Fits. This British Standard document recommends the correct limits for each part once the designer has decided the nominal size of the component and what type of fit and machining methods are to be used.

BS 4500 is long and complicated but it is important to be aware of its existence.

It is worthy of note that for items with small tolerances there is a higher production cost; therefore designers only put small tolerances on components when it is absolutely necessary to do so.

EXERCISE 3.3

From the drawing of the three-dimensioned features on a pulley assembly, examine the dimensions carefully and complete the table below.

Pulley Assembly

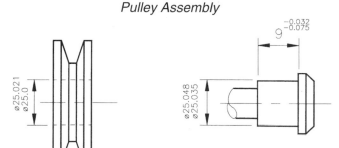

(a)   (b)

	Pulley hole	*Shaft length*	*Shaft diameter*
Nominal size:			
High limit:			
Low limit:			
Tolerance:			
Type of fit of shaft in pulley:			

Witness testimony

The above exercise has been completed satisfactorily by . . . .

Signed . . . . Job title . . . . Date . . . .

EXERCISE 3.4

1. In the drawing of a **cast iron bracket** opposite, measure the drawing and add all dimensions and centre lines in accordance with BS 308. Remember that the drawing is to a scale of 2:1 (i.e. the finished component will be two times bigger than that shown). Use the indicated faces as datums. Indicate on your drawing that the bore on the large hole must be within +0.10 and −0.05 mm of the nominal size.

2. Complete the title block by adding:

   • your initials
   • the appropriate projection symbol
   • today's date.

Witness testimony

The above exercise has been completed satisfactorily by . . . .

Signed . . . . Job title . . . . Date . . . .

BS 308 drawing conventions

Symbols and abbreviations are used extensively on engineering drawings; they save space, preventing drawings from becoming cluttered. It is essential that you understand these symbols and abbreviations or that you know where to refer to in order to find their meaning.

Some of the most common symbols and abbreviations used in engineering drawings are found in the following British Standard publications:

BS 308 Engineering drawing practice
BS 499 Welding terms and symbols
BS 970 Specifications of steels
BS 1134 Methods for assessment of surface texture
BS 2917 Specification for graphical symbols used on diagrams for fluid power systems and components
BS 3939 Graphical symbols for electrical power, telecommunications and electronics diagrams.

EXERCISE 3.5

Task
Refer to BS 308 or other appropriate source of information and complete the sketches in the right-hand column to show the BS convention for the illustrated features. In each case, state the source of the information used and how it was confirmed that the information is up to date. The convention for an external screw thread is shown completed in the first box for your guidance.

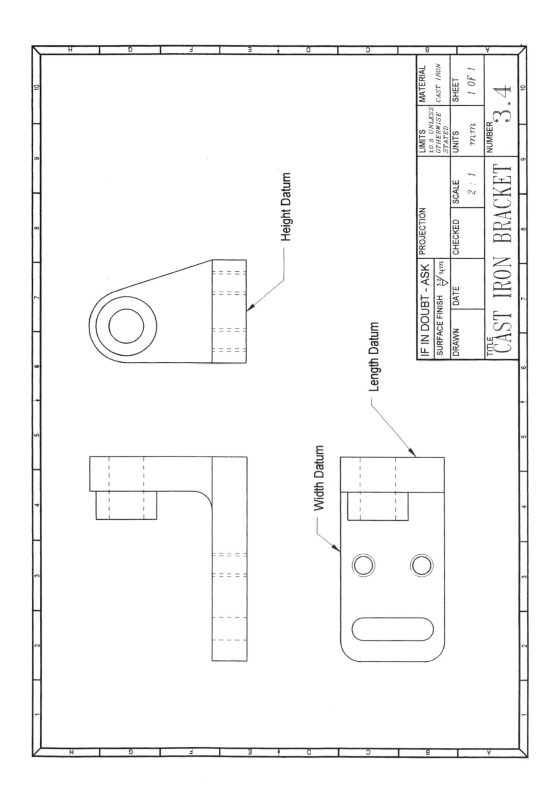

Height Datum

Length Datum

Width Datum

IF IN DOUBT - ASK

SURFACE FINISH 3.2 √m

PROJECTION

LIMITS
±0.5 UNLESS
OTHERWISE
STATED

MATERIAL
CAST IRON

| DRAWN | DATE | CHECKED | SCALE 2 : 1 | UNITS mm | SHEET 1 OF 1 |

TITLE CAST IRON BRACKET

NUMBER 3.4

Title	Feature	BS convention	Information source
External screw thread			
Internal screw thread			
Diamond knurling			
Ball & roller bearings			
Square on a shaft			
Flat on a shaft			
Holes on a circular pitch			

Witness testimony

The above exercise has been completed satisfactorily by . . . .

Signed . . . . Job title . . . . Date . . . .

BS 308 drawing abbreviations

EXERCISE 3.6

Task

Below is a crossword puzzle relating to abbreviations used on engineering drawings. Examine the clues carefully and then complete the crossword puzzle (you may need to refer to the abbreviations section in BS 308 Part 1).

Clues Across:
1. Chamfered
6. Maximum metal condition
7. Spot face
9. Undercut
10. Pneumatic
11. Counter bore
13. R
15. Spherical radius
17. Cylinder
19. Hydraulic
20. Material

Clues Down:
2. Head
3. Maximum metal condition
4. Pattern Number
5. Required
7. Standard
8.
11. Countersunk head
12. Right hand
14. Assembly
16. External
17. Centre line
18. Left hand

Witness testimony

The above exercise has been completed satisfactorily by . . . .

Signed . . . . Job title . . . . Date . . . .

EXERCISE 3.7

Task

Draw three views in **third angle projection** of the component shown. Use the paper provided overleaf. The grid should guide you in positioning the drawings correctly.

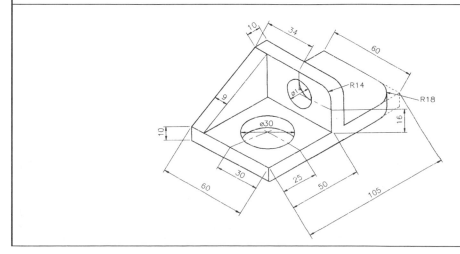

Witness testimony

The above exercise has been completed satisfactorily by . . . .

Signed . . . . Job title. . . . . Date . . . .

Sectioning

A sectional elevation may be of benefit when a drawing contains so much hidden detail that it has become unclear. A sectional elevation is a view of a part of a component when viewed on a cutting plane. The cutting plane on engineering drawings is shown as a fine chain line with thick ends, the viewing direction is indicated by two arrows against these ends with identifying letters, for example, A–A.

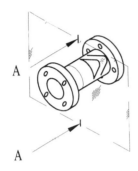

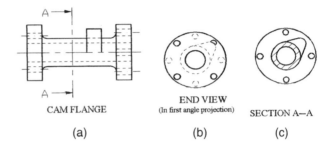

CAM FLANGE

END VIEW
(In first angle projection)

SECTION A–A

(a)　　　　　　(b)　　　　　(c)

Consider the cam flange shown above. When the end view is drawn in first angle projection, the resulting drawing contains much hidden detail. Compare this with the view Section A–A, which would be seen if the cam flange was viewed on the cutting plane (all material to the left of the cutting plane is ignored). The sectional view A–A shows more clearly the wall thickness of the tube, the shape of the cam and the positions of the holes in the right-hand flange. The area through which the cutting plane passes is hatched with fine lines drawn at 45°, usually about 4 mm apart.

The layout of a sectional view is always in accordance with the rules for orthographic projection. For example, the cam flange shown above is in first angle projection, that is, it is viewed in the direction of the arrows 'A', so the sectional view is drawn on the opposite side from where it is viewed.

Note. The features listed below are not hatched when a cutting plane passes lengthways through them:

- webs (supporting ribs)
- fasteners (nuts, bolts and washers etc.)
- shafts
- thin sheets.

When a component is sectioned through a hole or other similar feature, the hatching lines on all the areas of that component have the same pitch (spacing) and angle.

Half section

Used if a component is symmetrical, and both an assembly view and a sectional view would be of benefit.

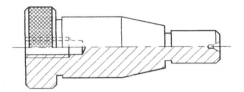

Sectional assembly

When drawing a sectional view through an assembly, it is correct to draw the direction of hatching lines reversed for adjacent parts. For clarification, the pitch of the hatching lines can be adjusted as long as the pitch remains constant for any single component.

Sectioned assembly looking on a pulley mounting bracket.

Note that the spindle, nut and washer are not hatched.

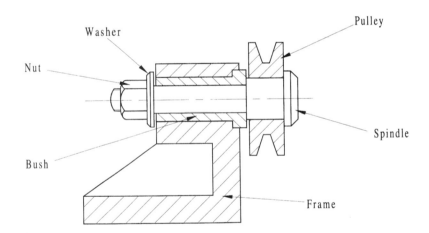

Other methods used to illustrate how assemblies fit together

- **Exploded views** – This system of drawing is sometimes used in service manuals to illustrate how complicated assemblies fit together. The chain lines indicate the axis of key components, for example shafts and locating pins.
- **General arrangement (GA) drawings** – A GA drawing shows how components fit together in an assembly and lists all the individual component drawing numbers. GA drawings do not show the components in detail nor do they give dimensions.

Exploded view of a Harrison M300 lathe slide's assembly Courtesy of the 600 Group

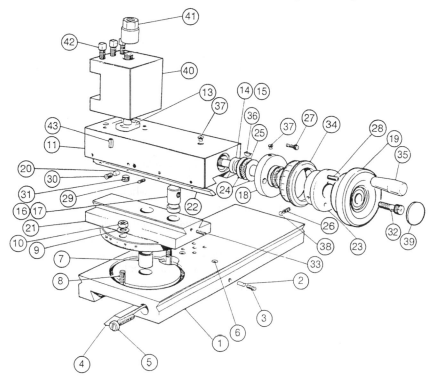

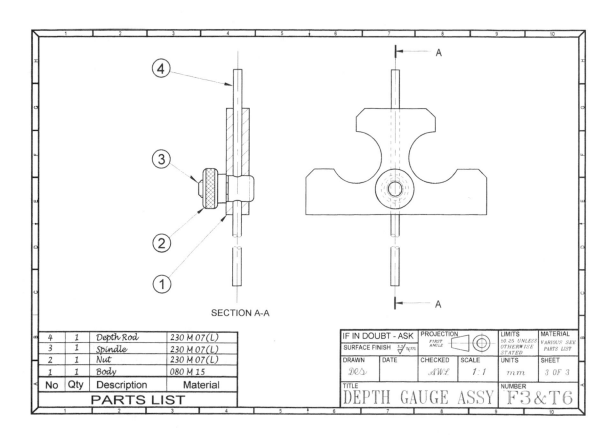

SECTION A-A

No	Qty	Description	Material
4	1	Depth Rod	230 M 07(L)
3	1	Spindle	230 M 07(L)
2	1	Nut	230 M 07(L)
1	1	Body	080 M 15

PARTS LIST

IF IN DOUBT - ASK	PROJECTION	LIMITS	MATERIAL
SURFACE FINISH	FIRST ANGLE	±0.25 UNLESS OTHERWISE STATED	VARIOUS SEE PARTS LIST

DRAWN	DATE	CHECKED	SCALE	UNITS	SHEET
DES		AWL	1:1	mm	3 OF 3

TITLE	NUMBER
DEPTH GAUGE ASSY	F3 & T6

Pictorial projection

A good way of communicating the shape of a component and presenting an image of how things fit together is to produce a pictorial drawing. Pictorial drawings are produced principally for technical illustrations and are used in manuals and sales catalogues of components. All the features of a component are not shown on this type of drawing as it shows a three-dimensional (3D) type image as it might be seen in real life. The back, one side and the underneath are all 'out of sight' and not shown or represented in any way. The two forms of pictorial projection are isometric projection and oblique projection.

Isometric projection

Isometric projection is used when a pictorial shape of the finished object is required. This type of projection shows a '3D'-like image, which is easier to visualise than an orthographic drawing. It is used extensively in books and catalogues as it can be understood by nonengineers. Isometric drawings are not usually used for production, so dimensions are not always included.

On isometric drawings, **all** horizontal lines on the X and Y planes are drawn at 30° and vertical lines are drawn vertically. All lines are drawn to a constant scale. An example of a component in isometric projection is shown.

Oblique projection

Oblique projection is used when curves and circles dominate the component, making it difficult to sketch in isometric projection. In oblique projection, the component's main front (usually the face containing the radii and/or circles) is drawn full size and true shape. The remaining faces are then filled in, receding at an angle of usually 45°.

(a)

(b)

(c)

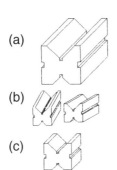

However, oblique drawings sometimes show the back faces receding at 30° or 60° to show different parts of the component in greater detail (see drawings).

For a more life-like appearance, the receding lines can be drawn half their true length.

EXERCISE 3.8   Production of an isometric sketch on isometric drawing paper

Task

Shown opposite is an angle plate drawn in first angle projection. You must sketch the angle plate on the next page in isometric projection.

The page has an isometric grid for your guidance, which should help you in proportioning the drawing correctly. Use a scale of 1 space on the grid to represent $\frac{1}{4}''$ of the angle plate (4:1). Take care to start in a suitable place on the page so as not to run out of paper.

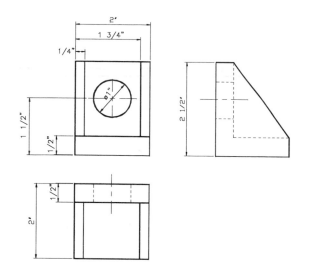

Witness testimony

The above exercise has been completed satisfactorily by . . . .

Signed . . . . Job title . . . . Date . . . .

EXERCISE 3.9   Production of an isometric sketch of a component

Task

On the blank page provided on page 95, sketch the milling clamp shown below. Your sketch must be in isometric projection and in good proportion.

Dimension your sketch in millimetre units so that it can be clearly understood.

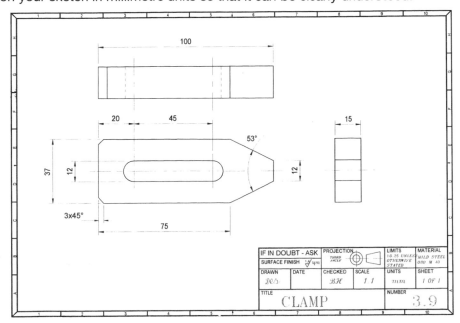

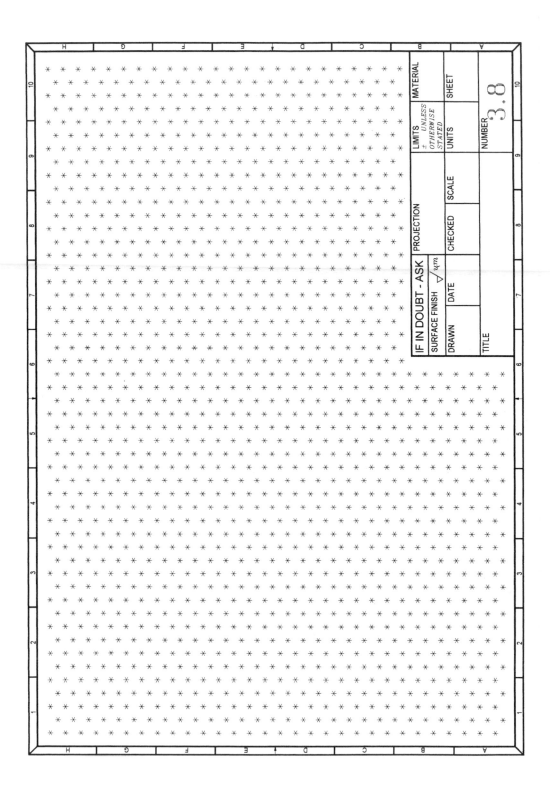

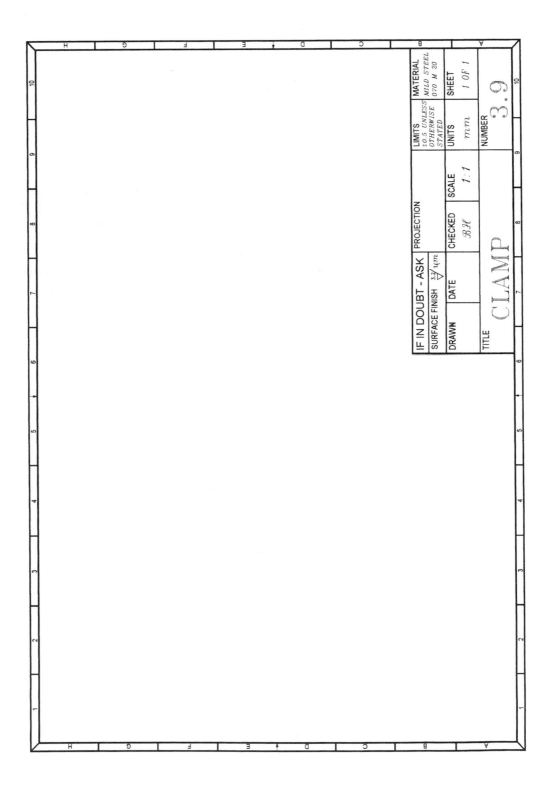

IF IN DOUBT - ASK

SURFACE FINISH ∇ 3.2 μm

DRAWN	DATE	CHECKED	SCALE
		B.H.	*1 : 1*

PROJECTION

LIMITS	MATERIAL
±0.5 UNLESS OTHERWISE STATED	*MILD STEEL* 070 M 20

UNITS	SHEET
mm	*1 OF 1*

NUMBER

TITLE CLAMP

3.9

Witness testimony

The above exercise has been completed satisfactorily by . . . .

Signed . . . . Job title . . . . Date . . . .

Sheet metal developments

When a sheet metal component is being designed, the designer has to convey the shape of the sheet metal **before** it is bent and folded to the sheet metal worker. For this purpose, a drawing called a *development* is produced. A few sheet metal developments are illustrated below, together with a pictorial view of the shape that each would assume after folding to give an appreciation of what a development drawing looks like.

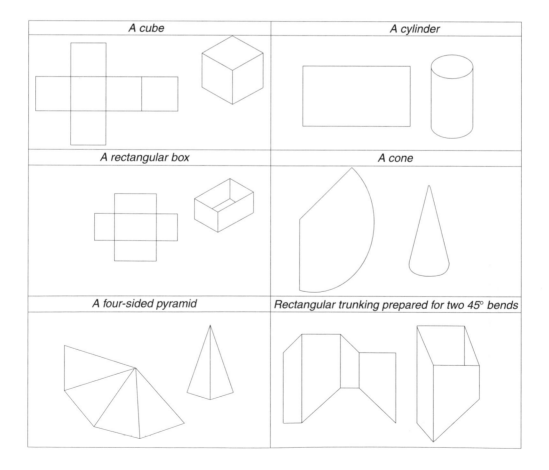

A cube	A cylinder
A rectangular box	A cone
A four-sided pyramid	Rectangular trunking prepared for two 45° bends

EXERCISE 3.10

1. Sketch the shape of the development that would be drawn for the following sheet components:

(a) Tool box	(b) Computer case	(c) Two pieces that make up a funnel

2. Make the following sheet components through paper developments.

(a) A cover for this book	(b) A paper wallet for three standard $3\frac{1}{2}$ inch floppy discs

Witness testimony

The above exercise has been completed satisfactorily by ....

Signed . . . . Job title . . . . Date . . . .

4 Communicating technical information in the workplace

Exercise checklist

(Ask your assessor to initial the lower box for each exercise when you have completed it.)

Exercise no.	4.1	4.2	4.3	4.4	4.5	4.6
Initials						
Date						
Exercise no.	4.7	4.8	4.9	4.10	4.11	4.12
Initials						
Date						

An important part of an engineer's job is to be able to communicate effectively. Communication takes a range of forms, but whatever form is being used, the message needs to come across clearly so that there is no doubt as to what the message is.

Imagine you are on holiday, abroad, on a lovely sunny beach. But your team is playing a vital European match. You have a portable radio, a lounger – everything you need to help you enjoy your listening. The match starts but the reception is poor. There is a great deal of crackle and you are finding it difficult to follow the match. But it is only the first leg and you will be home for the second leg. You decide to abandon the radio and go and play the match for real on the beach with your friends.

Supposing what you had been trying to listen to had been the final, would you have given up so easily?

Look at the picture. Does the commentator at the match know what is happening to you? Can you do anything to let the BBC know that you cannot hear what is being said? Fill in the exercise below to help you analyse this form of communication.

EXERCISE 4.1

Is the communication one-way or two-way?	
What was the real problem that stopped you from listening to the match? Was it: • the commentator • your radio • bad reception • something else?	
If the match had been the final, explain why you might have made more effort to continue listening.	

Witness testimony

The above exercise has been completed satisfactorily by . . . .

Signed . . . . Job title . . . . Date . . . .

Communication within organisations

All organisations need to communicate with the people in the organisation and there are various ways in which this can be done; the following may be used in your workplace:

- Verbal instructions and information given out by supervisors or other line managers.
- Written instructions. These may be a variety of forms and could include requests for information from the Human Resource Manager (personnel), right through to Process Information, which is the written sequence of operations and notes for production.
- Bulletins and newspapers. Many firms have their own in-house newspaper or bulletins, which may be pinned to notice boards.

Electronic mail is becoming increasingly common but suffers from the same problems as other forms of communication in that there needs to be a check to ensure that everyone has read the information.

Effective communication

As we have seen, to be effective, communication needs to be a two-way process. All the ways of communication outlined above fall into the category of one-way information and hence have limited use. In engineering, it is essential that instructions are not only issued but are also understood because if not there might be safety implications. Most children would have played the 'whispering game' in which one person whispers something to another and the whispering goes on to the end of the line where the last person repeats what was heard. What this person says usually causes everyone else to laugh as the message has been corrupted along its path. Effective communication ensures that the person receiving the message receives the correct message and is able to ask questions, query information or generally feels free to discuss what has been received.

In the section on 'You and your Supervisor', the importance of you being able to talk freely to your supervisor, to ask questions and generally feel at ease with the person charged with your training was stressed. The same principle should apply to other forms of communication.

Look at the following exercises – the first asks you to identify the types of communication methods used in your workplace, while the second asks you to think about how to ensure that communication is effective.

EXERCISE 4.2

Identifying the types of communication methods present at work

Communication method	Used in your workplace? Yes or No	Notes to include advantages and disadvantages of method
Notices	*Yes*	*Pinned up in canteen. E.g. health and safety. May not be read by everyone.*

Verbal instructions		
Written instructions		
Written requests for information		
Firm newspaper		
Electronic mail		
Bulletin board		
Notice board		
Sketches		

Witness testimony

The above exercise has been completed satisfactorily by. . . . .

Signed . . . . Job title . . . . Date . . . .

EXERCISE 4.3

Thinking about effective communication

Communication method	One-way or two-way	Used by you in the following way:
Discussion with supervisor	*Two-way*	*To discuss progress, to clarify points I do not understand.*
Company newspaper		
Written instructions		
Request to senior colleague for help		
Request to tool room for equipment		
Letter of application for a job		
Telephone call to a company for technical assistance, e.g. COSHH information		

Witness testimony

The above exercise has been completed satisfactorily by............................

Signed . . . . Job title . . . . Date . . . .

Communicating technical information

Technical information can be very difficult to pass on using only words. A great deal of engineering information has to be very accurate, so it is much easier to use drawings and symbols with clearly defined and understood meanings. When different materials are specified, it is important to know the exact specification. Here, we use British Standards that can be looked up on data sheets. If someone goes to the stores, the store person may also consult data sheets or data held on computer to track down the exact component that is requested for.

EXERCISE 4.4

Witness testimony

This is to confirm that.............................has explained satisfactorily why information is presented in different ways.

Signed . . . . Job title . . . . Date . . . .

In the section on engineering drawing, you have been told how drawings are an ideal way of communicating technical information but that drawings have to be made using their own 'language', which is understood by those who have been trained to make engineering drawings. What you also need to know is that technical information can quickly get 'out of date'.

(a)   (b)

Suppose you were sent to find drawings for details of wheel nuts. You look through the drawings, find wheel nuts but fail to look at the date! Drawings have **currency**, which means that only current drawings should be used. Drawings also have to be **valid**. It is possible that a range of prototype parts were produced using different materials. These were then tested and the best specification was chosen. This specification would then be the **valid** specification and no other specification should be used. Obviously, if the wrong drawings are used, it could be both expensive and dangerous, so firms need to have rules to ensure that only valid and current drawings are issued.

EXERCISE 4.5

You have been instructed to make some 'spares'. What would you do to make sure that the drawing you obtained was the correct one to work from to make the spares needed.

Witness testimony

The above answer shows that the candidate understands the importance of working from drawings, which are issued in such a way that they are current and valid.

Signed . . . .   Date . . . .

Most people are familiar with the specialist guides that help you to service and repair cars. These guides are valid for some models but not for others. Also, for people working in the service sector, there will be price guides with up-to-date information.

Another area in which information needs to be up to date is **process information.** This is information that sets out how a whole process is undertaken. Such information can take a great deal of time and effort to put into place. Process information is generally held in the area where the process is being carried out. The problem with this sort of information is that because it takes such a long time to sort out the effort required to review it, when alterations to the process are made, it may not be put into the process information folder. Quite often, new ways of doing things are passed on by word of mouth and this can make life difficult for a trainee.

Sketches

What you may find is that your supervisor explains the new way of working by using a sketch. Sketches are basically freehand drawings that lack accuracy but can be very useful as a means of communicating information. You might also use a sketch to give someone directions. Sketches can give a 3D view and can also be used to solve problems. Many an engineering problem has been solved using a sketch on a beer mat!

So, if you are in doubt as to what you should be doing, check the process information.

If the information appears to be out of date, use common sense and initiative to decide what you should be doing. If you are in any doubt or there is any concern that what you think is the right answer might be wrong or could cause damage, check with your supervisor.

Other sources of information

In addition to the sources of information outlined above, there are a number of other sources of information.

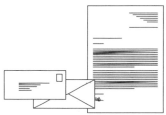

- **Books** – For general enquiries, an encylopaedia or a dictionary can produce the information required. Increasingly, people access these via the internet, but one should be aware that information on the internet is only as good as the person who wrote it. Information in books and magazines is checked thoroughly before being published and the author identified.

  Textbooks and maintenance manuals can also be very useful, but do check when they were published. Like drawings, they become out of date.
- **Mail** – The post is a method of sending and receiving information.
- **Publicity brochures and sales catalogues** – These are easily acquired, both nationally and internationally. They are an excellent way of getting to know about new products and their availability. You may need to request price guides as prices, being subject to change, are often not included in the publicity information.
- **Counsellors, data sheets and wall charts** – These are often supplied free to customers. A counsellor is a small pocket-sized booklet containing essential information such as conversion tables.

Letters

Publicity brochures

British Standards Kitemark

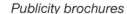

- **Telephone and fax** – These provide a quick way to request and supply information, e.g. control of substances hazardous to health (COSHH) data sheets.
- **BSI publications** – These offer the most reliable source of detailed information on recommended standards for quality. By having national standards, quality and consistency can be achieved. If drawings are to BSI engineering drawing practice standards then there is no confusion as to what symbols mean. Typical standards used will be:

  BS 308 Engineering drawing practice

  BS 499 Welding terms and symbols

  BS 970 Specification of wrought steels

BS 4643 ISO metric screw threads.

- **CD-ROMs, computers and the internet** – Nowadays firms frequently send out catalogues in CD-ROM form and it is also possible to access company web sites for up-to-date information.
- **Signs and labels** – Many packages carry signs to show how they should be handled. Make sure you read these as they are present to protect both you and the contents. There may also be signs with instructions about work practices (e.g. switch off mains before removing cover).

Machines are often labelled with an identification number, which should be quoted when reporting a fault or breakdown. Labels can also be fitted to equipment, for example, a label fitted to a chain may show its safe working load (SWL)

- **Appendices** – Appendices such as the ones at the end of this book may provide extra information.

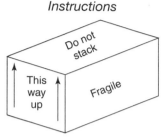

Instructions

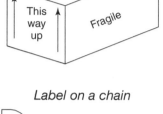

Label on a chain

Appendix

Appendix 1

Millimetre to inch conversions

mm	inch	mm	inch
0.01	0.0004	7	0.2756
0.02	0.0008	8	0.3150
0.03	0.0012	9	0.3543
0.04	0.0016	10	0.3937

Selecting appropriate sources of information

It takes practice to find, select and use the most relevant and appropriate information for your requirements. Information should satisfy the following criteria. It should be:

- **Clear** – The information should be clear, logical and straightforward to understand. If it is not, then try other information.
- **Suitable** – Choose a source of information best suited to what you want. For practical tasks, it may be more suitable to ask an experienced colleague rather than look in an instruction manual.
- **Relevant** – The information you find must be relevant. For example, if you want information on cutting speeds for a specified wrought steel, the best place to look is a table of cutting speeds, which can be found in a counsellor.
- **Understandable** – If you are not sure that you have understood the information, you should reread it or check with the person who has given you the information.
- **Up-to-date** – This is really important as specifications, standards and prices change and out-of-date information could cause problems.
- **Reliable** – If information has been translated from a foreign language, or if you have downloaded it from the internet, it may be wrong. Remember the case recently when NASA had done the original calculations for the Mars landing in inches and subsequent calculations were metric!

EXERCISE 4.6

Checking that information is relevant and up to date

Information required	*Details of information source*
The total stopping distance for a car travelling at 50 mph.	Relevant source of information: *Can be found in the Highway Code.* Date of issue: 1999 Date of expiry: N/A
The colour used to identify the live wire on a 13 A plug	Relevant source of information:
The recommended mileage between oil changes for a specified vehicle	Type of vehicle: Source of information: Date of issue: Date of expiry:
The pitch of a 14 mm motor vehicle spark plug thread	Source of information: Date of issue:
The decimal equivalent of 1/64 in.	Source of information: Date of issue:
The price of 100 m of 1 mm^2 three core flat with earth electrical cable	Source of information: Date of issue: Date of expiry:

Witness testimony

The above exercise has been completed satisfactorily by . . . .

Signed . . . . Job title . . . . Date . . . .

EXERCISE 4.7

Identify two people who are qualified to confirm that the information you have discovered is valid and reliable.	Name: Qualified because: Name: Qualified because:
Identify two people whom you have consulted to clarify information.	Name: Clarified information concerning: Name: Clarified information concerning:

Witness testimony

The above exercise has been completed satisfactorily by. . . . .

Signed . . . . Job title . . . . Date . . . .

EXERCISE 4.8

Witness testimony

I confirm that. . . . . has checked that information needed is valid, reliable and relevant.

Signed . . . .   Date . . . .

Information deficiencies

When working from drawings or from written information, you may sometimes find that the information appears to be deficient, that is, there appear to be some bits missing. When this is the case, it is best to look again and check to make sure you have not made a mistake.

If there is a deficiency in the information, it may be a genuine omission in the information or it could be that the missing information is routine to the persons doing the job regularly and that what you are looking for would not normally be shown. After thoroughly checking the information, if you still think there are deficiencies in the information, look at an alternative source to make up the deficiency. When you have found the missing information, ask your supervisor to check it.

Making up information deficiencies requires you to use your initiative. In some instances, the missing information may be obvious and a quick check will confirm that your guess was correct. In other instances, there may be several alternatives and you will need to consider whether making a decision is actually the right course of action. Your best action will be to check the alternatives with your supervisor and discuss with him/her which is the correct option, why and how you should proceed. Although we often learn from our mistakes, we also need to consider that making mistakes can be costly and should be avoided.

EXERCISE 4.9

Describe an example of information that you know is deficient but is readily made good.	
Describe an example in which you noticed information was deficient but needed to check which alternative was the correct answer.	

Witness testimony

I confirm that. . . . . has been observed to use his/her initiative and supply possible solutions to missing information. S/he has discussed problems and difficulties and has carried out agreed actions correctly.

Signed . . . .   Date . . . .

Communicating about yourself

What do I do or say to let people know what a great person I am? (Well I think I am great!) I suppose I communicate by:

- talking to people
- how I behave
- how I dress
- how I react and what my face tells other people
- how I write about myself.

The importance of dress and behaviour towards others has already been considered. Writing about yourself in a formal manner is an important skill. The Latin words **curriculum vitae** (CV) are used to outline the story of your life. What you need to do is to make sure that you communicate all the really important information about yourself in a manner that is easy to read and not too long. Look at the following exercise and decide which points are important:

EXERCISE 4.10

Look carefully at the information in the list that follows and organise the information into 4 columns. In the first column, put the information that is essential for a CV; in the second column, place information that may be useful but not essential; in the third, put information that could be useful; and in the last column, put any irrelevant information.

Name, football team supported, place of birth, educational qualifications, age, gender, religion, marital status, car driver with a clean licence, breakfast cereal preferred, medical details, courses attended that are work related, address, political affiliations, details of departments worked in and skills acquired, positions of responsibility achieved, out-of-work interests, names of people from whom a reference can be obtained.

Essential Information	Useful information	Additional information	Useless information

Now that you have sorted out the information, you need to organise it into a readable document. A CV can be handwritten but you should try to word-process your CV. Look at the example that follows and use it as a model. The layout and format can be personalised, but remember that although you want to stand out, you do not want to stand out for the wrong reasons. Your CV communicates about you in more ways than what is written on the paper.

CURRICULUM VITAE
for
Mustapha Skill

Address: 5 Link Road, Greenfold, Leicestershire.

Date of Birth: 2nd May 1984.

Status: Single, male.

Qualifications: 8 GCSEs: Maths (B), English Language (C), Science Double Award (BB), English Literature (D), French (D), CDT (C), History (D), Drama (E).

Clait level 2 Word Processing and Spreadsheets, level 1 databases. **NVQ2 Performing Engineering Operations** including key skills at Level 2.

Work: I have worked for 'Designed for You' since leaving school and have enjoyed this work and particularly enjoyed my time in the CAD section.

School achievements: I played cricket for the school Under 15 Eleven and am finishing my Duke of Edinburgh Bronze Award.

Work experience: In Year 10, I spent one week working for Joe's Auto Spares and enjoyed this work. In Year 11, I spent two weeks with the maintenance department of B. J. Smith and Sons who manufacture cam shafts for the motor industry. I have a Saturday job in a DIY shop in Greenfold and I help my brother run his disco.

Other interests: I like skate boarding and go with friends to local skate parks. I am helping my uncle renovate an old Morris car. I also like music and socialising.

EXERCISE 4.11

Develop and produce an up-to-date CV for yourself. The CV should be word-processed. (This is an opportunity to achieve key skills in IT. Either get your IT supervisor to check your on-screen document and then witness your carrying out a spell-check and on-screen editing, or, print an early version and then your final version showing the corrections you have made.)

Either write your CV in the space provided or stick in a word-processed version:

CV for . . . .

Witness testimony

I confirm that. . . . . .completed the following CV satisfactorily.

Signed . . . . .Job title . . . . Date . . . .

Give you a reference? Of course. Good luck.

The main use for a CV is when you are applying for jobs, and the best advice is that you should rewrite your CV for each job you apply for. By doing this, you think carefully about what you are writing and take the time to emphasise skills and experiences that you think are important in this particular case. A photocopied CV for another job may indicate to a would-be employer that you cannot always be bothered to make much effort.

Finally, do remember that if you send off a CV for a job and it includes the names of people who have offered to give you a reference, it is only polite to inform these people that someone may be in touch with them asking for a reference for you.

Exercise 4.12 is designed to help you achieve key skills in communication and also to increase your knowledge about an engineering topic of your choice by carrying out some research and then presenting this information to an audience by means of a short illustrated talk.

This assignment allows you to achieve a number of key skills both in communications and IT. One of the skills needed is that you are able to read and extract information from a variety of sources. The main task here is that you have to give a short talk, but to prepare for the talk you will need to do some research.

Reading and summarising information

Reading is still one of the main ways by which we gain information. It can be a quick and efficient way of finding out what we need to know. In addition to reading, we often need to summarise the main points of what we have read. We may need to do this for our own use or to pass information to someone else. For example, you may read an operation manual for a new machine. Then someone may ask you what a particular button or lever does. You summarise what you have learnt by telling them. (You could of course tell them to read the manual themselves, but this would not be very efficient.)

Occasionally, we need more than one point of view or we may need very up-to-date information; hence, we need to refer to more than one source of information. Sometimes, to our horror, we find that the information we require is part of a very long and, perhaps, boring article; thus, we need to develop the skill of skimming through the information and looking for keywords. Once we have found the keywords, we can read the actual information we want quite quickly. For example, you may find that Health and Safety documents are rather long and you may want to find out only whether suppliers have obligations under the Act. So you skim through the paragraphs quickly until you come to the word 'suppliers' and then you read in more detail.

When you read, you may come across very strong words such as 'you must' or 'it is your duty'. These words set the tone of the article. The writer is implying that you should take real notice of what is being said as it could affect you.

Writers also use 'signal' words such as 'therefore', which indicates that if you have done one thing then another thing logically follows. For example, 'a lathe is used to remove metal, **therefore** the piece of metal is smaller at the end than at the beginning!' Another such word is 'whereas', which indicates that you could do one thing or that you could do something else.

Images are another way in which meaning can be stressed. Most engineers would rather have a diagram or a flow chart than lots of words.

So to summarise:

- Reading is a useful way to gain information.
- Information gained from reading can be summarised.
- Gaining information from more than one source enables us to get up-to-date information and consider different points of view.
- Long articles can be skimmed and keywords identified.
- Writers use 'signal' words to stress meaning.
- Images often enhance meaning.

EXERCISE 4.12   Reading and summarising information assessment

You must read and summarise information from **two** extended documents about a straightforward subject. One of the documents should include at least **one** image.

Reading summarised from the following two documents:

1.
2.

What you need to know and do	Assessor initials	What the evidence consists of. (Possible evidence suggestions in italics.) Assessor can highlight or add additional comments.
Select and read relevant material		*Material selected may be one or more articles used in the preparation for the short talk, or articles may be given for a purpose, e.g. to find out about COSHH.*
Material should be from different sources		*Information could be taken from a range of sources, e.g. technical manuals, Health and Safety leaflets and the Internet.*
Identify accurately the lines of reasoning and main points from text and images by skimming articles that should be at least three-textbook		*Notes from articles used to prepare information for the short talk.* *Formal summaries done as a class exercise.*

pages long, and identify the main points of the article, either in writing or verbally, to your assessor. You should also be able to identify why an image has been used within an article. Also, you should be able to identify from the tone of writing and the use of 'signal' words what stress the writer is putting on various parts of the article.		*Verbal response to questions on information that is skimmed.* *Written response to information that has been skimmed.* *Verbal or written comment on why an image (such as a diagram, chart, flow diagram etc.) has been used.* *Verbal or written comment on the tone of an article, e.g. an article about health and safety where obligations are quite clearly stressed.*
Summarise the information to suit your purpose.		*Summary for the short talk.* *Summary at the end of a given article.* *Summary to another trainee.* *Summary could be written or verbal but the language used should be appropriate to suit the purpose of the summary.*

Witness testimony

This confirms that. . . . . .has demonstrated satisfactorily that s/he can read and summarise information.

Signed . . . . Job title . . . . Date . . . .

Giving a short talk

Task 1

Choose a topic that interests you and that your supervisor approves. Research this topic and **note down** from where you got all the information, for example, web site, technical manual, magazine. Write these references down in the space provided in the assessment box below. Try and use at least two web site references.

Task 2

Decide the title of your presentation (e.g. *Crash helmet design and manufacture*).

Task 3

Sort out the information into a logical sequence, which should have a clear **introduction** in which you indicate the things you are going to talk about, **followed by** several sections in which you describe and explain matter. You should **end** with an overview of what you intend to say.

Task 4

The presentation needs to contain at least one image. So think about how you are going to illustrate your talk. You may be providing facts and figures, so a graph or chart, or some pictures of different designs, would be useful.

Task 5

You want your talk to be remembered, so produce a word-processed handout, which should also contain a copy of at least one of the images you have decided to use. Check that your handout is clear and that there are no spelling, punctuation or grammar mistakes. **Either ask** someone to watch you do this checking **or** print out the incorrect version and then the corrected version.

Task 6

Get all your information together, read through the section on 'Giving a Short Talk', practise giving the talk and make sure it is long enough (about 5 minutes). Make sure you know enough about the subject to be able to answer questions. **Give the talk** and ask for questions at the end.

There are some additional notes to help you prepare your talk at the end of this chapter.

To help you with this preparation, use the following checklist and fill in all the information requested.

Things to check	*Give date when final check was made or enter the required information*
The title of my talk is:	
I have used the following as sources of information:	*www.* *www.*
I have decided to use the following images to make my talk clearer:	

I have sorted out my presentation and checked that it has a clear introduction and several sections that follow logically from my introduction and a conclusion.	Give date on which you did final check.
I have read through the notes on giving a talk and practised my presentation. *(I do not want to have to give another one because I made too many mistakes!!)*	
I have produced a word-processed handout that contains at least one image and got someone to check my use of spell-check/or printed out an uncorrected version.	
I have thought about some possible questions I could be asked and know the answers.	
I am ready to give my talk.	

Short task assessment

You must give a short talk about a straightforward subject using an image.
(This form to be completed by assessor at the end of your talk.)

Topic:	Date:
Audience:	Location:

What you need to know and do	*Assessor initials*	*What the evidence consists of. (Possible evidence suggestions in italics.) Assessor can highlight or add additional comments.*
Prepare for the talk by carrying out research and by noting the references used to include at least two web sites.		*Research notes* *Presentation plan* *Information on what will be covered.*
When you give your talk, you should speak clearly and in a way that suits your subject, purpose and situation.		*Clear speech with not too many 'umms', 'arrs' and pauses.* *Technical terms explained. Standard English used.*

Keep to the subject and help your listeners by giving pointers as you go along.		*A well-structured talk with a clear beginning, middle and end.* *New points identified.* *Logical sequencing.*
Use at least one image to illustrate your talk.		*Chart, diagram, model, drawing, real parts used.* *Reference made to the image.*
Produce a word-processed handout that contains at least one image.		*Handout produced that contained at least one image.*
Invite questions so that you listen to and respond to what others say.		

Witness testimony

This confirms that............................has given a short talk and that the criteria outlined above have been achieved.

Signed . . . . Job title . . . . Date . . . .

N.B. If you allow questioning at the end you may start a discussion and as a result obtain evidence for taking part in a **discussion**. The tracking sheet for this can be found in Chapter 2.

Key skills – communication level 2

Giving a short talk

Giving a short talk on a topic you know about is often called 'giving a presentation'. Many employers rate this skill very highly because it is a useful way of communicating information in the workplace. You may be called upon to find out about new software, or perhaps you have been sent on to a short course to learn to operate some new machinery that is to be installed in your workplace. When you come back to work, you may be asked to explain matters to your colleagues or show them how to use the new machinery or software. This is where knowing how to give a presentation comes in.

Hints on giving a good talk

- Be prepared. Make sure you know what you are talking about and be ready to answer questions.

- Remember that by the end of your talk you want people to have learnt something. So, if you use technical words (jargon) that you learnt, then you need to explain what they mean. It is a good idea to write them somewhere for people to see, and this also helps them to remember.
- Practise your talk before you give it.
- Do not talk too quickly.
- Look at your audience, make sure they are paying attention. You may need to say something like 'OK, let's get started' or 'Can I have your attention please'. While you give your talk, try and look around and make sure you include everyone. Looking at people's faces will tell you a lot about how well your talk is being received.
- Make sure that your talk is logical, that it has a beginning in which you outline what the talk is about; a middle where you do most of the explaining and an end where you summarise what you have said and again stress the main points that you have made.
- Provide illustrations, charts, diagrams, and so forth. People learn far more when they can see things as well as hear them.
- At the end of the presentation, be prepared to answer questions. If you do not know the answer to something, own up! Then tell the person that you will try and find out the answer.
- Give the talk in a location that suits you, for example in the workshop or in a computer room, so that you can demonstrate the use of software.

Things to avoid

- Do not be disorganised and unprepared. Remember, you could be wasting people's time.
- Do not rush. If you have been asked to give a presentation, then do your best to do a good job. At the end, you do not want people to say 'Well, that was a waste of time.'
- Do not use bad language.
- Do not give the impression that you feel that the presentation is not important.
- Do not be embarrassed. We all have to speak up in front of others on occasions, for example at a meeting or even a wedding!

5 Identifying and selecting engineering materials

Exercise checklist

(Ask your assessor to initial the lower box for each exercise when you have completed it.)

Exercise no.	5.1	5.2	5.3	5.4	5.5	5.6	5.7
Initials							
Date							

Atoms are the smallest particles of any substance that can be found. There are about 90 elements occurring naturally on earth and all materials are made from elements. A material that consists of only one element is said to be pure; however, most materials in their natural state are mixtures of many elements and are known as raw materials. These raw materials must be processed if they are to be of use in industry.

Processed materials are of most interest to engineers. These are those materials that are reasonably plentiful and have properties suited to tooling and manufacturing. There are two basic types of material used extensively in engineering: **metals** and **non-metals**; if a material is composed of two different materials bonded together it is known as a **composite.**

Metals

A metal is a raw material that is mined from the ground as an ore. After being solidified from the molten state, metals develop a crystalline structure and become good electrical conductors. Metals are not generally used in their pure state but are mixed together to form metallic mixtures called **alloys**. This mixing, or alloying, is to enhance certain properties or to make an alloy that is easier to work, that is to cut, bend or shape.

Metals can be divided into two categories: **ferrous** and **non-ferrous**. Ferrous metals and alloys contain the element iron. Steels are alloys of iron and small amounts of other elements, particularly carbon and manganese.

Ferrous metals

Plain carbon steels are the most versatile and widely used ferrous metals. They contain small amounts of manganese, and are usually identified by their carbon content (between 0.1 and 1.5%). The medium and high-carbon steels can be hardened and tempered (toughened) all steels can be softened (annealed). See the chart below for some uses of steels:

Composition of plain carbon steels		
Common name	*% carbon*	*Common uses*
Mild steel (or low-carbon steel)	0.1–0.25	General purpose, used throughout engineering
Medium carbon steel	0.25–0.7	Hammers and bolts and high stress components
High-carbon steel (or tool steel)	0.7–1.5	Metal cutting and forming tools

Mild steel
Sheet metal tool box

Medium carbon steel
Lifting hooks and chains

High carbon steel
Hand cutting tools

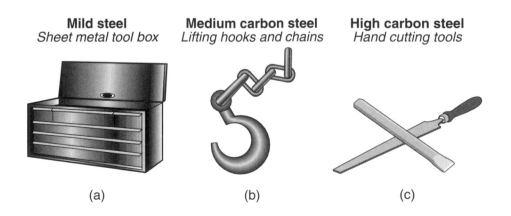

(a)   (b)   (c)

Cast Iron is iron mixed with much larger amounts of carbon than the plain carbon steels. Grey cast iron is the most common type of cast iron, and has a carbon content of 3.5% together with other elements in small proportions. Cast iron is used because it is very fluid when molten, which enables both large and intricate castings to be produced. Cast iron is quite brittle and has the quality of self-lubrication. The excess carbon in its structure causes both these properties. Grey cast iron is used extensively for the production of pre-machined parts, for example motor car engines and machine tool frames. It is also used for marking out equipment, for example vee blocks and angle plates.

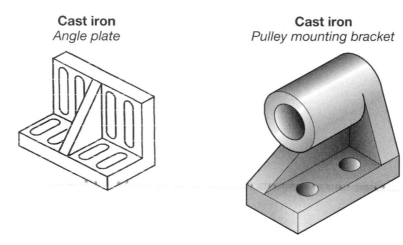

Cast iron
Angle plate

Cast iron
Pulley mounting bracket

Alloy steels are steels that have metals other than plain carbon steels in their composition. Alloy steels frequently contain such elements as chromium, nickel, molybdenum and vanadium. They are usually specially developed for particular purposes, for example stainless steel is particularly resistant to corrosion and is consequently used in the food processing and chemical industry. High-speed steel (HSS) is a special alloy steel developed for its hardness and toughness and is used for the manufacture of metal cutting tools, for example twist drills and milling cutters.

Stainless steel
Domestic sink unit

High-speed steel
Milling cutter

Non-ferrous metals

Non-ferrous metals are those metals and alloys that do not contain iron. Aluminium, copper, tin, lead and zinc are a few examples of pure non-ferrous metals. Each metal has its own properties and uses, and although copper is used for electric wires and water pipes, in general, non-ferrous metals are mixed together to form alloys. Many non-ferrous metals and alloys are corrosion resistant.

The chart below shows the composition of some common non-ferrous alloys used in engineering:

Metals alloyed	Common composition	Name of alloy
Aluminium	96%	Duralumin
Copper	4%	
Copper	60%	Cast brass
Zinc	40%	(Muntz metal)
Copper	97%	Coin bronze
Zinc	2½%	
Tin	½%	
Copper	94%	Phosphor
Tin	5%	bronze
Phosphorus	1%	
Tin	30%	Tinmans
Lead	70%	solder
Tin	60%	Electrical
Lead	40%	solder

Duralumin
Aluminium step ladder

Cast brass
Pipe fittings

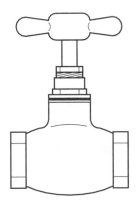

Coin bronze
1p and 2p coins

**Phosphor
bronze**
*Bearing
bush*

Tinmans solder
Stick for sheet steel solder

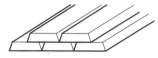

Electrical solder
*Solder for electronic
circuits*

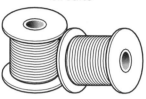

Non-metals

Non-metals are diverse in their makeup. They can be:

- **plastic** – for example thermosetting/thermoplastics
- **organic** – derived from plants, for example rubber and wood
- **mineral** – for example oil, glass and silicon/stone
- **synthetic materials** – for example special types of rubber.

Plastics

Plastics represent a large group of non-metals. They are the most common type of non-metals used in engineering. Plastic materials tend to be:

- lightweight
- strong in relation to their weight
- good insulators
- resistant to corrosion from acids.

Plastic materials have been synthetically made by processing organic and/or mineral materials. They have giant molecules that bond together in different ways to produce two distinct types of materials. The two types of plastics are recognised by how they behave when heated:

Thermoplastics are materials that are usually moulded while hot as they can be reshaped by heating. The molecular structure is long chain or branched. Some common thermoplastics are perspex, nylon, polyvinyl chloride (PVC) and polythene. Some uses for thermoplastic materials are illustrated below:

Perspex
Safety glasses to BS 2092.2

Polyester
*Towing strop/car
seat belts*

PVC
Wire insulation

Nylon
Hose fittings

Thermosetting plastics are shaped in moulds, and when heated they harden by chemical action. Their molecular structure is cross-linked, which gives the material hardness and rigidity. Some common thermosetting plastics are epoxy resin, glass fibre and urea formaldehyde; common uses for these hard and rigid materials are illustrated below:

Epoxy resin
Adhesive

Glass fibre
Lightweight sports car body

Urea formaldehyde
Electrical components

Softwood pallet

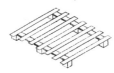

Hardwood exterior house door

Wood

Wood is a natural material, most types of wood are strong in tension along the grain but weak across the grain. Types of wood are:

- **Softwood**, which is derived from conifer trees, is used for pallets, crates, chipboard and paper manufacture. Softwood is also used for making the mould patterns used in the casting process.
- **Hardwoods** are the products of deciduous trees and tend to be stronger and more decorative than softwoods. Some examples of uses of hardwoods include hammer handles (ash), furniture frames (beech) and stairways and exterior doors (oak and mahogany). Experienced people can identify types of hardwood by colour and appearance of the grain.

Other types of wood include **manufactured boards,** such as plywood, chipboard and hardboard. These are cheaper and have the advantage of being available in large panels. Manufactured boards are used as follows:

Rubber

Known for its flexibility and durability, rubber is a natural product extracted from a particular type of tree. After it is processed (*see synthetics*), rubber is used for belts, tyres, antivibration mountings and drive couplings.

Plywood
Shelving

Chipboard
Kitchen cupboards

Perforated hardboard
Tool rack

Stone

Surface plates can be made from granite or glass as well as the more common cast iron. Glass is a very hard wearing and rigid material, which makes it suited to this application. Granite is a natural stone made up of mica and quartz and is a particularly hard-wearing material. Granite is found naturally in large pieces and can be machined into slabs. It is well suited for use as both large and small surface plates of high quality.

Granite surface plate

Ceramics

Materials having very good thermal resistance. They are a chemically bonded metal and non-metal combination and have a very rigid structure. Ceramics are used for grinding wheel grit, cutting tool tips and high-voltage insulators.

Synthetics

Synthetics are an artificial combination of materials resulting in a new material that possesses enhanced properties. A synthetic material is used for safety shoe soles. Various materials are added to rubber to make it tougher, more hard wearing and oil resilient.

Composites

This material group is made up of bonded components, which are combined to enhance each component's properties, usually its strength. Well-known composites include the following:

Composite	Content	Sketch
Reinforced concrete	Steel rods inserted into concrete structures – the composite has the compressive strength of concrete and the tensile strength of steel.	
Plywood	Sheets of thin wood glued together with their grain running at right angles to each other – the resulting sheet is much stronger than wood of the same thickness.	
Corrugated cardboard	Three layers of cardboard glued together – the resulting structure is light, stiff and has shock-absorbing qualities.	*Cardboard (cross section)*
Motor car timing belt	Synthetic rubber and nylon cord are combined to form a motor car timing belt. The rubber ensures silent running while the nylon cord provides strength.	

(GRP) Glass Reinforced Plastic	Thermosetting plastic and glass fibre laid in different directions result in a very strong and light material. It can be used for car bodies and motorcycle fairing and has the advantage of being corrosion resistant.

EXERCISE 5.1

1. Identify **two** examples of each material type listed below that is found in your workplace. Your examples should **not** have been quoted in the text of this book:

	Ferrous	*Non-ferrous*	*Non-metal*
(a)			
(b)			

2. Below is shown a diagram that illustrates the way in which engineering materials are classified.
Insert in the boxes a typical material in each category and state what it is likely to be used for.

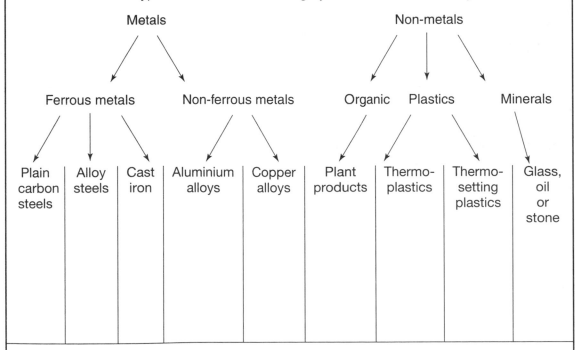

3. For each material type listed below:

- put a ✓ if the material has this characteristic
- put a ✘ if the material never has the characteristic
- put a **0** in the box if the characteristic is sometimes, but not always, present.

	Ferrous	Non-ferrous	Non-metal
Contains iron			
Flexible			
Is lightweight			
Electrical conductor			
Thermal insulator			
Alloyed with metals to improve properties			
Is magnetic			

A motorcar tyre is a composite material. State three advantages of this composite over a single material, when used as a tyre.	(a) (b) (c)

Witness testimony

I confirm that. . . . . knows the differences between materials that are natural/raw and materials that are metal alloys or synthetic materials.

Supervisor's signature . . . .   Date . . . .

Physical properties of materials

A designer chooses the material for a product and plans its manufacturing process. In order to do this, the designer must understand the properties of a wide range of materials.

The principal mechanical properties of materials, together with a simple workshop test where appropriate, are described below. Technical tests are available for most properties to enable accurate comparisons to be made if required.

The properties of materials, particularly metals, can be changed by alloying; for example, the addition of manganese to steel makes the steel easier to forge and roll. There are very few metals used in their pure state and most are alloyed with other elements to enhance the properties of the material or to make it easier to manufacture.

- **Hardness** – The degree to which a material can resist indentation, abrasion or scratching. Hard materials are difficult to cut either by sawing or filing, so these processes can be used as a simple test to compare different materials for hardness. For example, metal-cutting tools need to be hard because they cut other materials.

- **Toughness** – The degree to which a material can resist repeated hammering without permanent deformation. A simple toughness test can be devised by lightly hammering materials with the same force and by seeing which deforms the most. Hammerheads must be tough because they endure repeated blows.

- **Strength** – The degree to which a material resists various forces without breaking. There are three different types of strengths that are listed below:

  (a) *Tensile strength* – The property of withstanding tensile (stretching) loads without breaking. A simple test is to try to bend the material, if it is hard to bend, it has tensile strength. For example a bolt, when tightened, is in tension, so it must exhibit tensile strength.

  (b) *Compressive strength* – The property of withstanding compression loads without crumbling or breaking. For example, building bricks need to withstand the weight of the wall above.

  (c) *Shear strength* – The property of being able to resist cutting by a shearing action. A simple workshop test for shear strength is to try to cut pieces of different materials of the same thickness with combination pliers and then note which need the most force. For example, the hinge pin in a pair of scissors must be able to resist more shear force than the material being cut.

- **Ductility** – The degree to which a material can be permanently stretched when subjected to a tensile force. A simple test to compare the ductility of materials is to see how difficult it is to bend similar samples of each material. For example wire drawing needs a ductile material, so that it can be stretched into long fine lengths.

- **Conductivity** – A good conductor will allow either heat or electricity or both to pass through it. Most conductors allow both heat and electricity to pass through them. For example, cooking pans need to allow the heat from the cooker to get to the food inside.

- **Malleability** – The degree to which a material can be shaped by compressive loading, for example hammering. A workshop test for malleability is to squeeze samples in a vice and examine them to see which has deformed the most. For example, putty needs to be malleable to seal joints around house window frames because it is squeezed into the joint.

- **Corrosion resistance** – Corrosion resistance is the prevention of the wearing away or dissolving of materials due to environmental factors. In order to prevent carbon steels from corroding, a coating is usually applied to the exposed surfaces. This coating may be paint, plate or just a smear of oil or grease. Where the material is exposed to greater risk, special materials may be used, for example stainless steel is used in food-processing plants and corrosion of motorcar bodies is prevented by the use of hard paints.

- **Magnetism** – Only three common elements exhibit the property of magnetism, namely, iron, cobalt and nickel. A simple workshop test for a magnetic material is to see if the material is attracted to a magnet. Note that a material will appear to be magnetic if it has an electric current running through it. Copper is used as windings in electric motors because it is a good electrical conductor and can be made into a strong temporary magnet by having an electric current pass through it.

Screwdriver points may be magnetised to enable steel screws to be positioned in difficult corners.

- **Density/weight** – Each substance has a different weight for the same volume, and this is called the density. Density is usually measured in kg/m^3 (or g/cm^3). The density of some common materials is listed here:

Density	Aluminium	Brass	Cast Iron	Copper	Steel	Water
g/cm^3	2.72	8.48	7.2	8.79	7.82	1.00
kg/m^3	2720	8480	7200	8790	7820	1000

This means that one cubic cm of aluminium weighs 2.72 g, one cubic metre of aluminium weighs 2720 kg (or 2.72 tonnes).

- **Colour** – All materials can be identified by colour. Colour and hue (shiney or dull) can be used in a number of ways to identify materials:

(a) The outward appearance of a material changes from one material to another (gold and silver being the most obvious).
(b) The temperature of a metal can be estimated by its colour (steels are cherry red at $760\,^{\circ}C$).
(c) The colour of sparks from a grinding operation differs from straw colour to red depending on the type of material being ground (mild steel sparks are more yellow than high-carbon steels).
(d) Different materials put into a naked flame change the colour of the flame, for example copper changes a flame's colour to green).
(e) For wood, both the colour and grain structure can be used to identify the actual type.

Coins are made from different coloured metals, so that they can be quickly and easily identified.

- **Brittleness** – If a material is brittle it will not deform on breaking. Brittle materials are easily identified after being broken because there is little or no deformation of the pieces, and they can be reassembled (brittleness is not normally considered to be a desirable property). For example glass, cast iron and pottery are all brittle; after breaking them the pieces can be reassembled.

EXERCISE 5.2

1. Give one further example and indicate where and why each property listed below is required: Your example must not be taken from the text of this book. Include in your answers at least **one** of each of the following material types:
Plain carbon steel; alloy steel; thermoplastic; thermosetting plastic; mineral.

	Example	*Where used*	*Why suited*
(a) Hardness			
(b) Toughness			
(c) Compressive strength			
(d) Tensile strength			
(e) Shear strength			
(f) Ductility			
(g) Malleability			
(h) Corrosion resistance			
(i) Magnetism			
(j) Density			
(k) Colour			

2. Take the samples of marked materials from the tool store and conduct tests on them to find the required properties in the table. Complete the table by scaling your results from '1' to '6' in the following way:

(a) For hardness, write: '1' for the hardest and '6' for the softest.
(b) For toughness, write: '1' for the toughest and '6' for the least tough.
(c) For tensile strength, write: '1' for the strongest and '6' for the least strong.
(d) For shear strength, write: '1' for the strongest and '6' for the weakest.
(e) For malleability, write: '1' for the most and '6' for the least malleable.
(f) For colour, write: 'grey', 'silver', 'silver/grey', 'brown', 'golden/yellow' or 'silver/white'.
(g) For magnetism, insert either 'yes' or 'no'.

	Aluminium	Brass	Cast-iron	Copper	High-carbon steel	Mild steel
Hardness						
Toughness						
Tensile strength						
Shear strength						
Malleability						
Colour						
Magnetism						

Witness testimony

The above exercise has been completed satisfactorily by............................

Signed................Job title.............................Date..........

Changing the properties

The physical properties of many materials can be changed by simply heating and cooling. There are four basic processes used to change the properties of steels, and the various different types of steels respond in different ways to this heat treatment.

- **Annealing** – The process of heating up to an annealing temperature and allowing to cool as slowly as possible. Annealing leaves the steel in a softer state, making it easier to machine. Annealing temperatures depend on the percentage of carbon in the steel: mild steel (0.2% carbon) is annealed at 850 °C; medium carbon steel (0.5% carbon) is annealed at 810 °C; and high-carbon steel (1.0 % carbon) is annealed at 780 °C.
- **Normalising** – Heating as before, but allowing to cool in air at room temperature. Normalising returns the steel to its normal state and undoes the effect of any heat treatment.
- **Hardening** – Only medium and high-carbon steels can be hardened. To harden these steels, they must be heated to the annealing temperature and cooled quickly, usually by quenching it in water or oil. Hardening leaves the steel rather brittle and vulnerable to cracking.
- **Tempering** – The tempering process removes some of the brittleness that results from hardening. It allows the steel to retain some hardness and become more tough. The hardened steel is reheated to a precise temperature in the range of 230–320 °C, depending on the use to which the steel is to be put, that is level of toughness required, is then allowed to cool. The actual temperature of the steel being toughened can be judged by the colour of its surface while heating. This is only apparent if it has been polished before heating.

Colour	Temperature	Use
Pale straw	230°C	Metal-cutting tools
Dark straw	250°C	Punches
Brown	275°C	Chisels
Purple	300°C	Hammer-heads
Blue	320°C	Springs

EXERCISE 5.3

Take the box of eight material samples from the tool store. Examine and record each sample's properties in the table after conducting simple workshop tests and observations. From the information gained, you should be able to identify each of the material types.

The eight materials are listed as follows (in alphabetic order)

Aluminium	Brass	Bronze	Cast iron	Copper	High-carbon steel	Mild steel	Stainless steel

No.	Material colour (grey, silver/grey, yellow, white, golden brown, silver)	Magnetic (Yes / No)	Density (heavy, medium, light)	Hardness (hard, medium, soft)	I think this material is:
1.					
2.					
3.					
4.					
5.					
6.					
7.					
8.					

Write down a simple method of detecting if a piece of steel is mild steel, which contains less than 3% carbon, or medium/high carbon steel, which contains more than 5% carbon.	
Write down a method you could use to cool a piece of steel that is being annealed so that the cooling rate **is as slow as possible**.	
When sharpening a chisel on a grinding wheel, why is it important to keep the point cool by dipping it into coolant or water at regular intervals?	

Witness testimony

The above exercise has been completed satisfactorily by.............................

Signed . . . . .Job title . . . . Date . . . .

Forms of supply of materials

Black bar wrought iron gates

Hot rolled roof supports

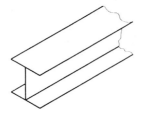

Ship's hull made from plate

Round section bar shaft

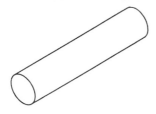

Rolled sheet steel for motorcar body panels

Raw materials can be supplied in many different forms. The choice enables companies to order their materials in a condition that will enable them to undertake their processing more efficiently. This results in savings to be made because the material is bought in the most appropriate form. It is therefore important to know and identify the forms in which materials can be supplied.

- **Black rolled bar** – Black Bar is supplied in long rectangular, square, hexagonal or round sections that are not smooth or precisely accurate in size. It is a cheap source of steel that is easy to cut. Black bar is usually selected when large amounts of machining are to be carried out on the workpiece, or the roughness of the unmachined surfaces is unimportant.

- **Hot rolled sections** – Often used for fabrications and in the construction industry, hot rolled sections are available in a wide variety of shapes and sizes. 'H', 'I' and 'Channel' sections are common. Angle bar is probably the most widely used hot rolled section but special sections can be bought, for example, railway lines.

- **Plate** – Plate is the term used for pieces of hot rolled steels. Plate is not smooth on its external surface, it is between 5 mm and 50 mm thick and up to 5 m wide. Plate is used in shipbuilding and boiler making.

- **Bright rolled bar** – Bright bar, as it is commonly known, is smooth and accurate in its dimensions. It is supplied in long lengths of constant cross section, usually square, rectangular or round. Bright bar is used if the external dimensions of the bar are similar to those of the finished product. It is not suitable for applications involving a lot of machining because it has internal stresses and is tougher than black bar (see above).

- **Sheet** – Sheet is supplied in rectangular cut lengths to most companies and is also available in coiled rolls for users of large amounts, for example the motor industry. Sheet has a smooth surface and can be bent, folded or pressed into a wide variety of products. Sheet is cold rolled, resulting in better dimensional accuracy and surface finish than plate. While most sheets are supplied at a thickness of around 1 mm to 2 mm, the thinnest mild steel sheet available is only 0.15 mm thick. Sheets can be supplied coated with zinc (galvanised), tin (tin plate) or plastic (colour coat) surfaces for corrosion resistance at extra cost.

- **Castings** – For intricate, complicatedly shaped components, it is common practice to purchase pre-made cast blanks of the component which are then ready for finishing. Sand casting is the most common method of manufacturing large iron items, whereas the die-casting process is used for smaller non-ferrous components that need a smooth surface.

Cast Iron vice frame (not the jaws)

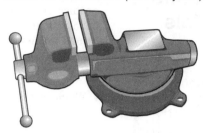

- **Forgings** – A forged component is one that has been shaped by hammering while hot. Forging is frequently selected as a manufacturing process, as it produces components with an enhanced grain structure, resulting in greater toughness and strength. Forging is also a quick method of producing a complicated shape.

Chisel and hammer head

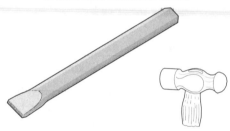

Electrical architrave

Used for many plastic products, e.g. computer parts

- **Extrusions** – Extrusion is to squeeze a material through a shaped hole. Toothpaste is extruded through the tube nozzle. Malleable metals and plastics can be extruded to make long lengths of regular cross-section strips for applications such as greenhouse frames and uPVC window frames. Pipes and tubes may well be extruded.
- **Mouldings** – This is the term used for some plastic shaping processes. Plastics can be die-cast in a similar way to metals. As polymers become plastic when heated, plastic sheets can be shaped around a pattern with air pressure (or a vacuum) to form products with very thin walls, for example chocolate box separators and plastic bottles.
- **Wood**
  (a) Wood can be supplied as rough sawn planks or as machined sections. Machined wood is usually about 3–5 mm smaller than specified, as it is measured before planing. Planed, tongued and grooved, dowel and mouldings are types of machined wood.
  (b) Laminated plywood, chipboard, hardboard or fibreboard are all available as wooden sheets. These are supplied as large sheets (2240 mm × 1220 mm in size). Veneered coatings can be applied to make the product better looking, stronger or waterproof. A special type of plywood called marine ply is more resistant to water but is expensive.
- **Copper-coated circuit board** – Electronic circuit boards are made from a Bakelite board that is coated with copper. The electronic circuit is etched onto the copper and the excess is removed in an acid bath, leaving only the copper tracks on the board's coating.

EXERCISE 5.4

Look around the workshop and find the products listed in the table below.

- For each product, state the probable form of supply to the original manufacturer.
- State the reason why the selected form of supply was chosen for the product.

The machine frame has been filled in for your guidance.

Product	Form of supply	Reason for using this form of supply
Machine frame	*Casting*	*Cast iron is rigid and strong*
Steel rule		
Screwdriver handle		
Filing cabinet		
Ladder		
Vice jaw		
Socket screw head		
Surface plate		
Workshop bench or table		
Pliers		

Witness testimony

The above exercise has been completed satisfactorily by...............................

Signed . . . . Job title . . . . Date . . . .

Defects in materials

On delivery, before any material is accepted, it is good practice to visually examine the materials for defects. Some common defects are described below, together with some simple methods of detection. (Other more sophisticated methods are mentioned where appropriate). The effect on the serviceability of the materials with the defects is also introduced.

- **Cracks** – If a bright rolled steel bar contains excessive sulphur it is prone to cracking. Forged components can crack if they have

Cracks are common in heat treated components such as this chisel

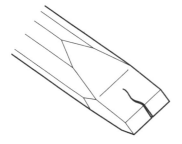

been hammered while not hot enough. Cracking can also occur if a component has been cooled too quickly during heat treatment.

Cracks weaken materials and should be avoided in all highly stressed components.

A reliable method of testing for cracks is to spray a penetrating dye over the surface of the workpiece being tested. When the dye has soaked into the surface of the component, it is wiped dry, and a developer powder is dusted onto the component's surface. If a crack exists, the penetrating fluid is drawn out by the developer and shows as a stain. More reliable methods of crack detection are to use X-ray equipment or ultrasonic equipment.

- **Blowholes** – Blowholes are large voids inside castings and are caused by poor venting during manufacture.

Blowholes become exposed when a casting is machined

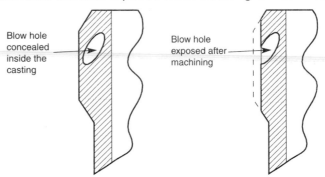

Blow hole concealed inside the casting

Blow hole exposed after machining

Blowholes cause weakness and are unsightly. Although they are invisible immediately after casting, a blowhole can become exposed during machining. The component would then be scrapped after time has been spent on its manufacture.

- **Slag inclusions** – This is associated with welded products or castings. Foreign matter would be found inside the component if it has slag inclusions. The foreign matter may be a flux from electric welding or some material that was oxidised during the casting process. Slag inclusions weaken welded joints, and as they are not visible, stressed components must be checked for this defect.
- To test for blowholes or slag inclusions, samples of the component can be sawn into pieces and visually inspected. If it is necessary to inspect all the components in a batch, X-ray or ultrasonic equipment would be used, as this would avoid the destruction of the castings.
- **Distortion** – Distortion takes the form of warping, bending and twisting. It is caused by:
  - machining one side more than the other side
  - machining without first stress-relieving the material
  - incorrect quenching method after heat treatment
  - uneven heating.

Distortion in bar stock can be detected by comparing the component's surface with a straight edge or a flat reference surface.

Distorted components can be straightened; this is time-consuming and might develop internal stress in the component.

- **Scale** – Scale is the term used for a hard area on the surface of cast or forged components. Scale can appear as flakes, which are

Distortion may occur during heat treatment

Loose scale should be removed with a wire brush

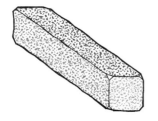

removed with a wire brush or file. If the scale is embedded in the skin of the component, it cannot be removed easily.

Scale can conceal small surface cracks and can also cause excessive wear on cutting tools.

Scale can be easily seen and should be removed, if possible, with a wire brush in order to expose any defects on the material's surface. In large steelworks, scale is removed with sulphuric acid.

- **Corrosion and oxidation** – If materials have been subjected to corrosive substances or kept out in the rain, the surface of the material is likely to be corroded. A common way of preventing corrosion is coating the surface of the material, often by painting or plating.

- **Timber knots and splits** – Wood that has dried faster at the ends may develop splits (called end shakes); if these splits are found, the ends should be cut off and discarded. Knots in wood may fall out or stain paint. In general, it is good to avoid knots in wood.

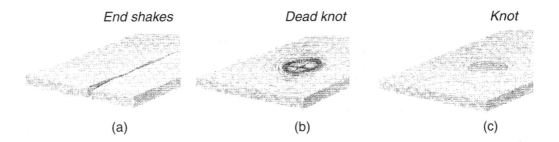

End shakes	Dead knot	Knot
(a)	(b)	(c)

- **Separation of laminates** – Many glued laminates are prone to separation if they have been wet. This can be experienced in the case of plywoods.

EXERCISE 5.5

1. List the defects you may encounter in (a) stock materials and (b) cast components.

(a) Stock materials	(b) Cast components

2. State two simple methods you can use to check that bar stock and sectioned are not distorted.

(a)	(b)

3. Give two causes of corrosion and oxidation on material surfaces.

(a)	(b)

4. State (a) why is stainless steel resistant to corrosion and (b) in what situations this property is utilised.

(a)	(b)

Witness testimony

The above exercise has been completed satisfactorily by. . . . .

Signed . . . . .Job title . . . . Date . . . .

'En' numbers and BS 970

During the Second World war, there was rapid movement in the development of steels. These new materials were given emergency numbers (En) for identification. In 1955, BS (British Standard) 970 was introduced to catalogue all materials. BS 970 was updated to enable easy identification of materials in 1983 and 1991; it is a four-part document and recommends specifications for wrought steels for mechanical and allied engineering purposes.

The section of most interest to general engineers is Part 1, which deals with inspection and testing procedures and specific requirements for carbon, carbon manganese, alloy and stainless steels.

The recommendation in BS 970 is to use a six-digit code to describe the steel's specification. The code is used as follows:

- The first three numbers represent the type of steel:
  000–199 indicates a plain carbon steel (the number is 100 times the manganese content).
  200–249 indicates a free-cutting carbon steel (the number is 100 times the sulphur content).
  250 indicates a particular type of silicon manganese spring steel.
  251–400 indicates a free-cutting alloy or stainless steel.
  300–499 indicates a particular type of stainless or valve steel.
  500–999 indicates a particular type of alloy steel.

- One of the following four letters follow to indicate:

  A – The steel supplied to meet chemical composition requirements.
  H – The steel supplied to meet hardening requirements.
  M – The steel supplied to meet mechanical property requirements.
  S – The steel is a type of stainless steel.

- The fifth and sixth numbers correspond to 100 times the amount of carbon in the steel.

An example of the BS 970 coding could be a steel of specification **BS 970: 070 M 26**.

This steel can be defined as follows:

(a) A manganese steel with 0.70% manganese content. (070 ÷ 100)
(b) Supplied on mechanical property specification. (Letter M)
(c) A carbon content of 0.26%. (26 ÷ 100)

Material colour codes and abbreviations

Most suppliers of steel paint a colour onto steel bars, enabling easy and quick identification of the actual material type. There is no BS recommended colour code for steels, so most companies have developed their own codes (the codes in the table below are used by a major British supplier).

Shown below is the BS 970 specification and old 'En' equivalent of some steels, together with the colour code used by a leading UK steel stockholder.

Material type	New BS 970 1991 specification	Old BS 970 1955 (En) specification	Common colour code
Low-carbon steel (or mild steel)	080 A 15	En3B (equivalent)	Blue
	070 M 20	En3B	Blue/red
	080 M15	En32B	Red
Free-cutting steels	230 M 07	En1A	Green
	230 M 07 (Leaded)	En1A (Leaded)	Magenta
Medium-carbon steel	080 M 40	En8	Yellow
Alloy steels	605 M 36	En16	White
	708 M 40	En19	Yellow/white
	817 M 40	En24	White/blue
	655 M 40	En36	White/red

Draughtsmen may use any of the following abbreviations to describe a material. Most of the following abbreviations are not specific to a material but describe the material in loose or general terms.

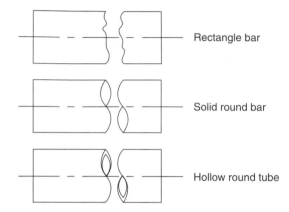

Rectangle bar

Solid round bar

Hollow round tube

MS = Mild steel
BMS = Bright mild steel
CI = Cast iron
HCS = High-carbon steel
Ally = Aluminium alloy

There are BS abbreviations for showing very long round bar, tube and rectangular sections and other features and components (*see also* BS 308).

EXERCISE 5.6

For the BS steel specifications listed, fill in the table below. The column headings are:

- the BS 970 specification for engineering materials (you can use the BS booklet or a supplier's catalogue for more information on this)
- common names of engineering materials
- the actual material composition of the listed material (you may use any source of information containing extracts of BS 970)
- your company's identification system for this material (usually a colour code).

BS 970 specification	Material type	Description of material composition	My company's identification system
1. 080 M 50	Medium carbon steel	Contains 0.8% manganese, supplied on mechanical property specification and has 0.5% carbon.	
2. 060 A 62			

3. 606 M 36			
4. 316 S 16			
5.	*High tensile steel*		

Witness testimony

The above exercise has been completed satisfactorily by.............................

Signed . . . . Job title . . . . Date . . . .

EXERCISE 5.7

When you start any job, you will normally be issued with the material to make it.

1. For each of the three given jobs listed below, state:

(a) The material from which the component is to be made
(b) Why this material was selected for this component
(c) The possible defects that could be present in the material you have been issued with.

2. State the checks you would need to undertake in order to:

(a) certify that you have been issued with the correct material
(b) ensure the material is free from defects.

	Fitting job: tool maker's clamp jaws	*Turning job: turning exercise*	*Milling job: turning tool holder*
Material	080M40	230M07(L)	070M20
(a) Properties that render this material suitable for the purpose			
(b) Possible defects	1. 2.	1. 2.	1. 2.

(c) How you would check to ensure that you are issued with the correct material type?			
(d) How to check for each defect listed in (b) above	1. 2.	1. 2.	1. 2.

List three different types of material normally used in the organisation in which you work. For each material type:

(a) describe the actual application of the product
(b) list the properties of the material that make it suitable for the purpose listed
(c) state an alternative material for the product.

Material types	Application	Properties	Alternative material
1.			
2.			
3.			

Witness testimony

The above exercise has been completed satisfactorily by.............................

Signed . . . . Job title . . . . Date . . . .

6 Checking workpieces for accuracy

Exercise checklist

(Ask your assessor to initial the lower box for each exercise when you have completed it.)

Exercise no.	6.1	6.2	6.3	6.4	6.5	6.6	6.7	6.8	6.9	6.10	6.11
Initials											
Date											

For the mechanical engineer, length is the most important measurement quantity. The standard unit of length in Great Britain, Europe and most of the world is the International Standards Organisation (ISO) **metric metre**. This unit is generally too large for engineering purposes, so it is divided into 1000 equal parts called ***millimetres*** (mm).

As many British companies have American connections, British engineers must also know about the imperial (inch) system of measurement. Imperial units are often stated as fractions of an inch (e.g. $^3/_4''$) or in decimals (e.g. 0.75″); 1″ is exactly 25.4 mm).

To avoid errors caused by expansion of materials due to temperature change, all precision measurements should be taken at a standard temperature of 20 °C.

A wide range of tools for measuring and marking out are available to engineers. After looking at the component's features to be measured or marked out, and considering the accuracy required, a decision on which tool is best suited to measuring is made. This decision is made with reference to:

- accuracy required
- linear or angular dimension
- access to the feature

- shape of feature
- size of feature

To ensure that appropriate readings are being given, it is necessary to use high-quality tools and to store and maintain them correctly. Keep your measuring and marking out tools safely in their original case and in a clean dry place when not in use. Before using them, it is good practice to check the accuracy of the instrument against a gauge of known size. If any error is suspected, it should be reported to the supervisor at the earliest opportunity to reduce any wastage that would otherwise be caused. When in use, tools should always be kept away from oil and dirt, reducing the possibility of their being damaged.

The most common measuring tools, together with their typical applications and methods to use them, are described below.

The engineer's rule

Engineer's rules are widely available in 150-mm (6 inch) and 300-mm (12 inch) lengths, although other lengths are available. Better quality rules are made from tempered stainless steel and have a non-glare satin chrome finish. Rules usually have both metric and imperial graduations engraved on alternate sides. The edges of rules are ground flat to allow straightness to be assessed. The finest millimetre division marked on metric scales is normally 0.5 mm; imperial scales are normally calibrated to 1/64″ (0.0156″). In practice, these graduations are very difficult to see, but rules are a quick way to measure components and are used extensively for making rough measurements.

A typical 150 mm (6″) engineer's rule

Tape rules

Roll-up tapes are used for measuring long lengths of up to a maximum of 30 m (100 ft). Measuring tapes are supplied with imperial and metric scales painted onto a steel blade. Shorter tapes of about 5 m (16 ft.) may be graduated in 0.5 mm (1/32″) intervals, whereas longer ones may be calibrated in 1 mm (1/16″) intervals only. It is difficult to read a tape over a long distance accurately as the tape may sag, expand or not be held straight.

Steel tape rules

Courtesy of Starrett

A tape rule is suited to checking such things as checking the length of steel barstock on delivery or for measuring the length of pipework systems.

EXERCISE 6.1 Measuring

Select and correctly use tapes and rules to check linear dimensions: hole sizes and hole positions within ±1 mm (or 1/32″).

Trainee to complete

Job title	
Job number	

I have used the rules/tapes to measure the workpiece features listed below and have listed my actual measurements below:

Measuring tool used	Feature measured	Size and limits	Measured size
300-mm (12″) steel rule			
150-mm (6″) steel rule			
Steel tape rule			

Witness testimony

I confirm that I have seen. . . . . .correctly select and use the above equipment to measure those features listed by him/her within the tolerances required by the drawing. . . . .

Signed . . . . Job title . . . . Date . . . .

Calipers

Calipers are tools that have two high-quality steel arms with rounded ends. Calipers are used to measure features that are inaccessible to other tools or to compare the size of one part of a component with other parts to find if the component is parallel. The distance between the ends of the arms can be adjusted to fit a component's size (either externally or internally). The size is then **transferred** to a suitable measuring tool (e.g. an engineer's rule or a micrometer) and a reading is taken. Calipers are non-indicating tools because they have no graduations to indicate the size. Because of the transfer of size, calipers cannot be as accurate as 'direct' measuring tools, but accuracies of about ±0.5 mm (1/64″) are readily achieved. With care and practice, more accurate readings

Spring-type external calipers

Courtesy of Moore and Wright

Firm joint-type internal calipers

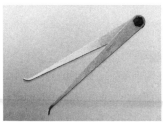

Courtesy of Moore and Wright

can be attained. Both internal and external calipers are made as either spring or firm joint type; they are set and used as shown below:

- **Spring-type calipers** – The legs of this type of caliper are opened or closed by means of the adjusting nut. They pivot on a roller and are tensioned by a bow spring. The external type shown is available in sizes from 75 mm (3″) to 300 mm (12″). Common uses for external calipers are measuring diameters of pipes, recessed bores and undercuts that would otherwise be inaccessible.
- **Firm joint-type calipers** – The caliper legs on the firm joint-type pivot on a large screw that incorporates fibre washers. These washers enable the settings to be retained. Adjustment for the size to be measured is made by opening or closing the legs by hand to the approximate size and then tapping one leg on a solid surface to make the final adjustment. The size of this type of caliper ranges from 150 mm (6″) to 600 mm (24″).

Method of transferring measurements with calipers

When transferring a measurement, the caliper legs are set to touch the work's maximum dimension. The size is then transferred to a rule or micrometer and checked as shown:

Set the caliper in the bore

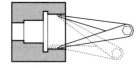

Read the bore size

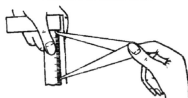

Set the caliper on a diameter

Courtesy of Moore and Wright

Read the diameter's size

EXERCISE 6.2

Select and correctly use calipers (both internal and external) to check linear dimensions: hole sizes and parallelism within ±1 mm (or 1/32″).

Trainee to complete

Job title	
Job number	

I have used calipers to measure the following features of my workpieces and have listed my actual measurements below:

Measuring tool used	Feature measured	Size and limits	Measured size
Internal calipers			
External calipers			

Witness testimony

I confirm that I have seen .... correctly select and use the above equipment to measure those features listed by him/her within the tolerances required by this exercise.

Signed .... Position .... Date ....

Engineer's squares

There are two types of engineer's squares used for assessing squareness, try squares and cylinder squares.

The try square

A try square is used on nearly all jobs for assessing the squareness of two adjacent edges. It is also used for marking out lines square to a side.

Try squares are made up of two parts – a thin parallel blade and a short thick base called the *stock*. The two pieces are assembled so that both edges of the blade are square to both edges of the stock.

Try squares are available in a range of blade lengths (from 75 mm to 600 mm) and grades to BS 939 (AA reference, A inspection and B workshop grades).

Your try square must be handled and stored carefully to retain its accuracy.

To check the squareness of a workpiece using a try square, hold the work against the stock of the square and slide the work towards the blade until contact is made. The assembly is then viewed against daylight to visually assess the size of any gap or error.

This procedure can be undertaken by hand or on a surface plate as shown below.

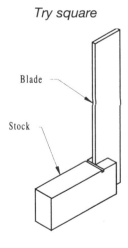

Try square

Blade

Stock

By hand with a try square

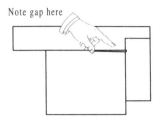

Note gap here

On a surface plate

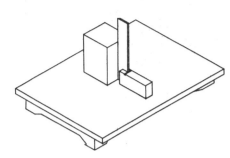

How a try square is checked for squareness is described on page 164.

A cylinder square on a surface plate

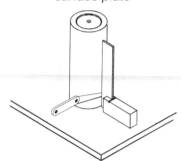

The cylinder square

Cylinder squares are case-hardened cylinders that have been ground so that the outside diameter is parallel. One end is ground so that it is square to the outside diameter. A typical cylinder square is about 300-mm high and ø75 mm. This type of square can be used on a surface plate for setting workpieces upright. Cylinder squares are sometimes preferred to try squares because they are more stable than try squares and are easier to make and restore.

Workshop protractor

Workshop protractors are instruments used for setting and measuring angles. They can be read to an accuracy of about ±1/2° (±30 minutes). Workshop protractors are usually supplied as part of a combination set. The components that make up a combination set are:

- hardened steel rule (usually 300 mm)
- moving protractor head that incorporates a spirit level
- square head with 45° edge and a spirit level
- centre head.

A combination set is shown below in some typical measuring situations:

(a) Assembled as a workshop protractor

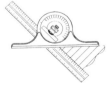

(b) Assembled as a centre finder

(c) Assembled as a try square

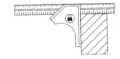

(d) Assembled as a depth gauge

To read a workshop protractor, first set the instrument so that the edge of the rule is in contact with one side of the work and the protractor head's measuring surface is in contact with the other surface of the work. The protractor scale is then locked in position and examined to see which angular division lines up with the zero line on the protractor head.

Although there is only one row of 180 calibrations (lines) on the moving protractor head, it is labelled in two directions (i.e. from 0° to 180° and from 180° to 0°).

By reading the most appropriate scale, the angle can be read as shown:

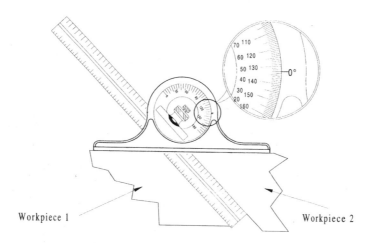

Workpiece 1 measures: 133°   *Workpiece 2 measures*: 47°

Practice readings:

Look at the two protractor settings below and write the indicated measurement in the space on the bottom line.

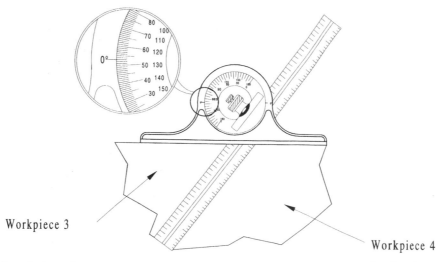

Angle of workpiece 1....   *Angle of workpiece 2....*

EXERCISE 6.3

Select and correctly use squares and protractors to check angles and squareness within ± 0.5 degree.

Trainee to complete

Job title	
Job number	

I have used squares and protractors to measure the following features of my workpieces and have listed my actual measurements below:

Type and size of protractor/square	Feature measured	Size and limits	Measured size

Witness testimony

I confirm that I have seen . . . . correctly select and use the above equipment to measure those features listed by him/her within the tolerances required by this exercise.

Signed . . . . Position . . . . Date . . . .

Profile gauges

Profile gauges are for assessing the accuracy of shapes and are manufactured with an accurate shape or profile. They are usually supplied in sets that cover a range of profiles. The most common profile gauges are:

- radius gauges
- screw pitch gauges
- point angle gauges.

(a) Radius gauges

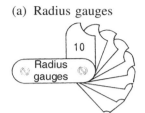

(b) Screw pitch gauges

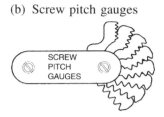

(c) Point angle gauges

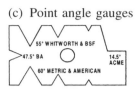

In use, each gauge is selected in turn from the appropriate set and compared to the work. The gauge is then held against the job and viewed with daylight in the background.

When using radius gauges, the differences in profile can easily be seen and the assessment of the quality of the work surface can be made.

(a) Work radius same as gauge

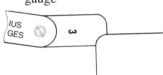

(b) Work radius greater than gauge

(c) Work radius smaller than gauge

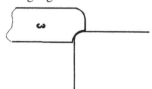

EXERCISE 6.4

Select and correctly use radius gauges and screw pitch gauges to check profiles within stated drawing limits.

Trainee to complete

Job title	
Job number	

I have used radius gauges and screw pitch gauges to measure the following features of my work pieces and have listed my actual measurements below:

Type of profile gauge	*Feature measured*	*Size and limits*	*Measured size*
Radius			
Screw pitch			
Point angle			

Witness testimony

I confirm that I have seen . . . . .correctly select and use the above equipment to measure those features listed by him/her within the tolerances required by this exercise.

Signed . . . . Position . . . . Date . . . .

Micrometer instruments

All micrometer instruments can measure features within 0.01 mm (0.001″), although some are more accurate. There are a number of different types of micrometers, the most common being the external fixed anvil type, which is used for measuring outside (external) dimensions. This type of micrometer has a rigid bow-shaped frame, a micrometer head and a fixed anvil (measuring face). External micrometers are available in a range of sizes and in either metric (mm) or imperial (inch) units. External micrometers are made in the following range of sizes:

$$0-25\,\text{mm}\ \ (0-1'')$$
$$25-50\,\text{mm}\ \ (1-2'')$$
$$50-75\,\text{mm}\ \ (2-3'')$$
$$75-100\,\text{mm}\ \ (3-4'')$$

and so on in 25 mm to a maximum of 450-mm intervals (18″).

The drawing below shows the names of the parts of a micrometer.

A standard 0–25 mm metric micrometer

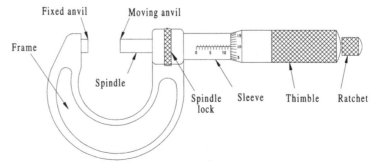

Micrometers are manufactured within small tolerances; they should always be carefully handled. Always ensure that the anvils are clean because appreciable errors can be caused by dirt and oil on the anvils. Clean by lightly squeezing a piece of paper between the anvils and sliding it out – the dirt or grease is then removed.

Setting and reading an external metric micrometer

(a) Identify the size of the micrometer required, noting the lower limit.
(b) Adjust the micrometer so that its anvils are both in contact with the component being measured, using the ratchet.
(c) Set the 'measuring pressure' with the ratchet.
(d) Count the number of whole millimetre divisions visible on the sleeve.
(e) Add on 0.5 mm if a $^1/_2$ mm is visible on the sleeve.
(f) Add 0.01 mm for each thimble division.
(g) Total the values of the above to get the reading.

Reading shown is 21.68 mm

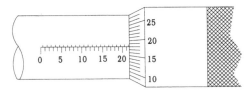

The above reading shows the following:

(a) The measuring range of the micrometer shown is 0–25 mm; the lower limit is therefore 0.
(b) There are 21 whole millimetres visible.
(c) A 0.5 mm is visible.
(d) The 18th division on the thimble lines up with zero on the sleeve.
(e) Total is 21.68 mm.

0
21.00
0.5
0.18
———
21.68
———

Setting and reading an external imperial micrometer

(a) Identify the size range of the micrometer required, noting the lower limit.
(b) Adjust the micrometer so that its anvils are both in contact with the component being measured.
(c) Set the 'measuring pressure' with the ratchet.
(d) Count how many 0.1″ divisions are visible on the sleeve.
(e) Count the quantity of 0.025″ divisions visible between the edge of the thimble and the last recorded 0.1″ mark and add their total.
(f) Add 0.001″ for each thimble division.
(g) Total the values of the above to get the reading.

1.000
0.075 (3 × 0.025)

0.017
———
1.592

Reading shown is 1.592″

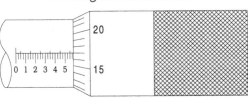

The above reading shows the following:

(a) The measuring range is 1–2″, so the lower limit of the range is 1.00″.
(b) There are five 0.1 divisions visible.
(c) There are three 0.025″ marks visible before the thimble edge.
(d) The 17th division on the thimble lines up with zero on the sleeve.
(e) Total is 1.592″.

Note If a micrometer's accuracy is suspected to be incorrect, it should be checked against a setting block and reset if necessary by a suitably qualified person.

Practice readings:

Look at the three metric and three imperial micrometer settings below and write the indicated measurement in the space on the bottom line.

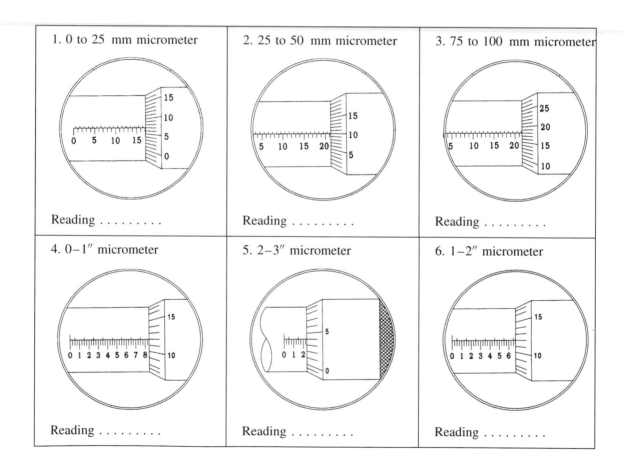

1. 0 to 25 mm micrometer

Reading . . . .

2. 25 to 50 mm micrometer

Reading . . . .

3. 75 to 100 mm micrometer

Reading . . . .

4. 0–1″ micrometer

Reading . . . .

5. 2–3″ micrometer

Reading . . . .

6. 1–2″ micrometer

Reading . . . .

Other types of micrometers

Apart from the usual external micrometer, other types are also available, although less common. These instruments are valuable accurate measuring tools. Three types are shown and described below.

An external adjustable micrometer with four anvils can measure from 0 to 100 mm

- **External adjustable micrometer** – This type of micrometer has an interchangeable fixed anvil that can be replaced with other precision anvils of varying length. This extends the measuring range of the instrument with minimal loss of accuracy. Reading to 0.01 mm (0.001″) is still possible. Adjustable micrometers are usually supplied with a set of precision checking gauges.
- **Depth micrometers** – These tools are used for measuring the depth of holes, slots and so forth. They consist of a hardened and ground base, a micrometer head and a set of interchangeable precision rods that can be used to give a wide measuring range, typically from 0 to 300 mm (0 to 12″).

Depth micrometer set containing hardened base, micrometer head and a range of interchangeable rods

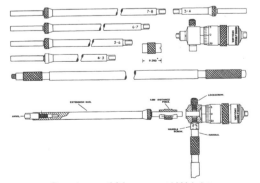

Courtesy of Starrett

Using a depth micrometer to measure the depth of a recess

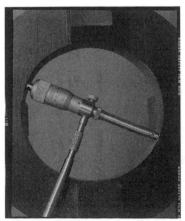

Courtesy of Starrett

- **Internal micrometers** – Designed to measure the internal dimensions of components. These instruments are also adjustable by means of a set of precision rods, giving a typical overall range of 50 to 200 mm (2 to 8″). The head on internal micrometers usually has a shorter measuring range than the other types of micrometers (13 mm or 1/2″). However, they can still measure to within 0.01 mm (0.001″) over their entire range.

An imperial internal micrometer set with measuring head, five precision rods, spacer and handle

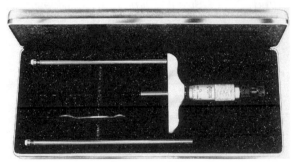

Courtesy of Moore and Wright

Internal micrometer in use

Courtesy of Moore and Wright

EXERCISE 6.5

Select and correctly use micrometers (external, internal, depth) to check linear dimensions: hole sizes and parallelism within ±0.025 mm (or 0.001″).

Trainee to complete

Job title	
Job number	

I have used............................to measure the following features of my workpieces and have listed my actual measurements below:

Type of micrometer	Feature measured	Size and limits	Measured size

Witness testimony

I confirm that I have seen.............................correctly select and use the above equipment to measure those features listed by him/her within the tolerances required by this exercise.

Signed . . . . Position . . . . Date . . . .

Vernier calipers

For slightly less precise measurement than the micrometer, a vernier instrument can be used. Vernier instruments have two engraved scales – a main scale and a vernier (sliding) scale. The vernier system allows fine measurements to be taken, in either millimetres or inches, by observing and reading the lines that coincide (line up) on the two scales. The main scale is calibrated in full-size divisions, whereas the vernier scale is calibrated in slightly smaller divisions.

Metric readings of 0.02 mm or 0.05 mm (depending on the scale) and imperial readings of 0.001″ can be reliably achieved by experienced engineers.

The most common vernier instrument is the vernier caliper. A vernier caliper is a useful tool as it has a long measuring range (150 mm, or 6″ is common) and a wide range of applications. It can be used for measuring the following measurements:

- External
- Internal
- Depth
- Step.

(a) Vernier caliper used for external shaft measurement

(b) Vernier caliper used for internal bore measurement

(c) Vernier caliper used for depth measurement

(d) Vernier caliper used for step measurement

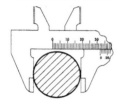
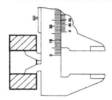
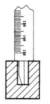
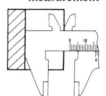

The vernier caliper shown below is of a typical design. The scales have fine black lines on a satin chrome background to enable easy and accurate reading. Scales are flush fitting, which reduces the chances of parallax (sighting) errors while reading. The jaws are fully hardened and their measuring surfaces accurately finished so that all measuring surfaces are parallel. The internal jaws are bevelled to reduce their contact area and to enable measurement of fine undercuts and similar features. The end of the instrument is square to the depth rod, which is thin and tapered at its end. Vernier instruments should always be carefully handled, kept clean and stored in their original cases.

Names of parts on a vernier caliper

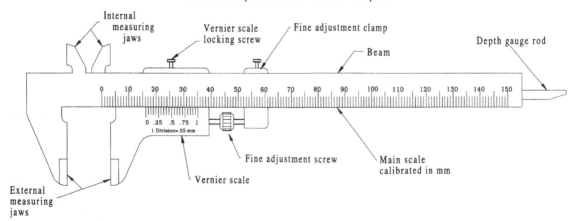

Precision vernier calipers are usually calibrated with a metric and an imperial scale; however, for clarity, the calipers illustrated in this chapter are shown only with millimetre units or inch units.

Setting and reading vernier calipers

(a) The caliper is roughly adjusted to fit the component.
(b) Lock the caliper in position with the small screw on the fine adjustment clamp.
(c) Turn the fine adjustment screw so that the anvils are a good fit on the workpiece.
(d) Lock the vernier scale into position with its locking screw.
(e) Read the measurement (see below for methods).

A vernier caliper set to read a metric scale (see details below)

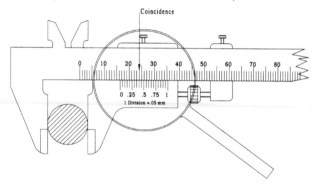

The method of reading each of the various vernier scales is slightly different. The procedure for reading each is described below.

Reading a 0.05-mm metric vernier scale

One of the two common metric scales (reading to 0.05 mm) is described here.

(a) Count the number of main scale divisions (usually whole mm) before the vernier scale zero line.
(b) Look on the vernier scale and observe the quantity of 0.05 mm divisions on the scale that coincides with (lines up with) any main scale division (it is marked * here).
(c) Add up the above values.

Reading shown is 16.40 mm

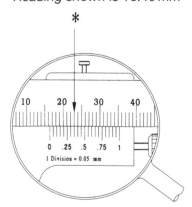

The metric (0.05 mm) example above is read as follows:

(a) There are 16 whole millimetres before the zero line on the vernier scale.
(b) The 8th vernier division lines up with a main scale division, each division representing 0.05 mm; so 8 × 0.05 = 0.40 mm.
(c) Total is 16.40 mm.

Reading a 0.02 mm metric vernier scale

Reading shown is 16.18 mm

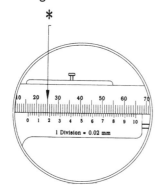

Another common metric scale (reading to 0.02 mm) is described here.

(a) Count the number of main scale divisions (usually whole mm) before the vernier scale zero line.
(b) Look at the vernier scale and observe the quantity of 0.02-mm divisions on the scale that coincides with (lines up with) any main scale division (it is marked * here).
(c) Add up the above values.

The metric (0.02 mm) example shows the following:

(a) There are 16 whole millimetres before the zero line on the vernier scale.
(b) The 9th vernier division lines up with a main scale division, each division representing 0.02 mm; so 9 × 0.02 = 0.18 mm.
(c) Total is 16.18 mm.

Reading an imperial vernier scale

Reading shown is 1.293

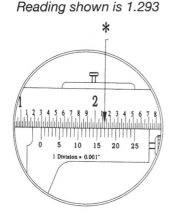

The most common imperial scale (reading to 0.001″) is described here.

(a) Count the number of inches before the zero line on the sliding (vernier) scale.
(b) Note how many 0.1″ are included between the vernier zero and the first whole inch mark.
(c) Note how many 0.025″ marks are included between the vernier zero and the last 0.1″ mark.
(d) Look at the vernier scale and observe the quantity of 0.001″ divisions on the scale that coincides with (lines up with) any main scale division (it is marked * here).
(e) Total all the above to get the reading.

The imperial example shows the following:

(a) One whole inch before the zero line on the vernier scale (1″).
(b) Two 0.1″ marks between the vernier zero and the 1″ mark (0.2″).
(c) Three 0.025″ lines between the vernier zero and the 0.2″ mark (0.075″).
(d) The 18th vernier division lines up with a main scale division, each vernier division representing 0.001″; so 18 × 0.001 = 0.018″
(e) Total is 1.293″.

Practice readings:
Look at the three metric and three imperial micrometer settings below and write the indicated measurement in the space on the bottom line.

1. 0.05 mm Vernier scale

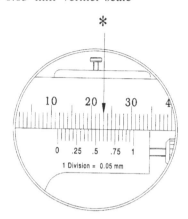

Reading . . . . . . mm

2. 0.05 mm Vernier scale

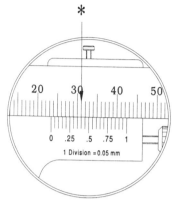

Reading . . . . . . mm

3. 0.02 mm Vernier scale

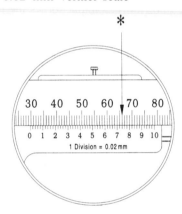

Reading . . . . . . mm

4. 0.05 mm Vernier scale

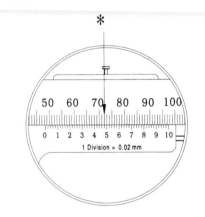

Reading . . . . . . mm

5. Imperial vernier scale

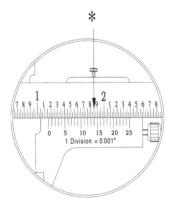

Reading . . . . ''

6. Imperial vernier scale

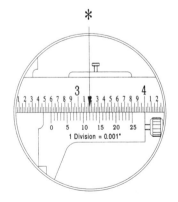

Reading . . . . ''

Calipers with digital read-out are available and are becoming quite common; these are more expensive than those described above but are much easier to read accurately in either inches or millimetres. The scale can be set at zero at any point enabling deviations from size to be easily read off. Digital caliper instruments are however slightly more bulky but have a lifelong battery and some have solar cells.

A digital caliper

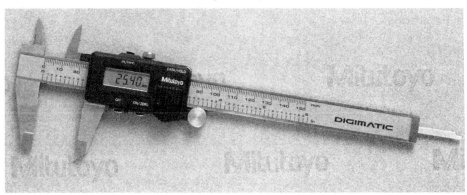

Courtesy of Mitutoyo

EXERCISE 6.6

Select and correctly use verniers (external, internal, depth, height) to check linear dimensions: hole sizes and parallelism within ±0.025 mm (or 0.001″).

Trainee to complete

Job title	
Job number	

I have used verniers (external, internal, depth, height) to measure the following features of my workpieces and have listed my actual measurements below:

Type of vernier	Feature measured	Size and limits	Measured size

State the advantage of using electronic measuring instruments:

Witness testimony

I confirm that I have seen . . . . correctly select and use the above equipment to measure those features listed by him/her within the tolerances required by this exercise.

Signed . . . . Position . . . . Date . . . .

Comparative measuring equipment

The measuring tools in the remainder of this chapter are **comparison** tools – they do **not** measure but compare the size or shape of known gauges to the component being made.

Feeler gauges

Feeler gauges are generally supplied as sets of shims (very thin pieces of high-quality steel). Each shim has a different thickness and is inserted into small gaps to assess the width of the gap.

Feeler gauges can be used to assess the amount of distortion in a steel bar after heat treatment by putting the work onto a flat surface and then inserting a thin feeler into a gap between the work and the flat surface. If the feeler gauge goes in, a thicker one is tried, the process is repeated until the thickest possible gauge or combination of gauges is found; this represents the size of the gap.

To check the accuracy of a try square, mount the cylinder square on a surface plate and hold the try square against the cylinder square. Feeler gauges can then be inserted into any gap to assess the try square's inaccuracy.

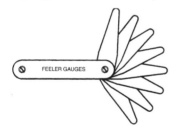

A typical set of metric feeler gauges (0 to 03–1 mm)

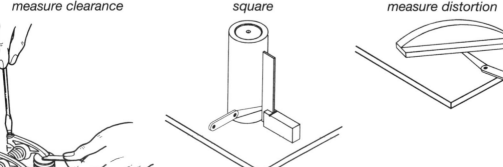

Using feeler gauges to measure clearance

Using feeler gauges to test a try square

Using feeler gauges to measure distortion

EXERCISE 6.7

Select and correctly use feeler gauges to check linear dimensions and flatness.

Trainee to complete

Job title	
Job number	

I have used feeler gauges to measure the following features of my workpieces and have listed my actual measurements below:

Feeler gauge thickness	Feature measured	Size and limits	Measured size

Witness testimony

I confirm that I have seen . . . . correctly select and use the above equipment to measure those features listed by him/her within the tolerances required by this exercise.

Signed . . . . Position . . . . Date . . . .

Limit gauges

When checking large batches of work to see if a dimension is within the limits, it is not considered economical to **measure** each component being inspected – it can be **gauged** more quickly. Gauging is a quick and easy method of determining whether a dimension is within the limits stated on the drawing; the process does not indicate the actual size of the component – it only indicates whether it is acceptable or not.

To test a component's features with a gauge, use a gauge that has been produced with two ends; one end is made to the maximum permissible size of the feature being tested and the other is made to the minimum permissible size.

Hole in workpiece to be inspected by a plug gauge

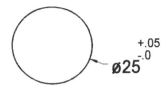

The plug gauge

This type of gauge is used to check if a **hole** is within the stated limits on the drawing.

The diameter of the hole shown must be between 25.0 mm and 25.05 mm. One end of the gauge is made to the minimum permissible size – it should go into the hole (else the hole is too small) and is called the **GO** gauge. The other end should not go into the hole (else the hole is oversize) – it is called the **NOT GO** gauge. (Note that the **GO** end is longer because it should enter the hole for its whole length.)

A typical plug gauge

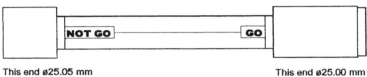

This end ø25.05 mm   This end ø25.00 mm

To test a hole, the inspector just tries each end of the gauge into the hole and ensures that the GO end goes in and the NOT GO end does not.

The result of testing a hole with a plug gauge could be as follows:

The GO end of the plug gauge will not enter the hole	The GO end will enter the hole and the NOT GO end will not enter the hole	If the NOT GO end of the plug gauge goes into the hole
The hole needs to be enlarged	The hole is within the limits	The component is scrap

Gap gauge

Another type of limit gauge is the gap gauge. The same principle applies, but the gap gauge checks a component's external features. Its **GO** end is therefore made to the minimum permissible size and the **NOT GO** end is made to the maximum permissible size.

Gap Gauge

Workpiece to be inspected with a gap gauge

To gauge the size of an external component, the inspector tries each side of the gauge onto the component in turn.

The result of testing a component's external feature with a gap gauge could be as follows:

The component will not enter the GO side of the gap gauge.	The component will enter the GO side of the gap gauge and not enter the NOT GO side of the component.	The component will enter the NOT GO side of the gauge.
The component is too big and needs to be reworked.	The component is within the limits.	The component is too small and is scrap.

EXERCISE 6.8

Select and correctly use plug and/or gap gauges to check linear dimensions: hole sizes are within stated drawing limits.

Trainee to complete

Job title	
Job number	

I have used limit gauges to measure the following features of my workpieces and have listed my actual measurements below:

Type and size of gauge	Feature measured	Size and limits	Measured size
Plug gauge			
Gap gauge			

Witness testimony

I confirm that I have seen. . . . . .correctly select and use the above equipment to measure those features listed by him/her within the tolerances required by this exercise.

Signed . . . . .Position . . . . .Date . . . .

Surface texture comparison plates

Surface texture comparison plates are sometimes called *scratch blocks*. They are generally supplied as sets of plates, each plate containing a number of samples of surface finishes produced by various machining methods to varying degrees of surface roughness. One plate may be samples of turned diameters, each surface representing a specified roughness average (Ra) value. The inspector selects the appropriate surface texture comparison plate and compares the roughness of the work to the various samples on the block by visual examination and by making comparative scratching with his fingernail. When the surface on the plate most similar to the work is found, its quality of surface finish can be read off as a number. The number is the Ra value, which is given in micrometres (μm).

Good-quality operations should result in the listed Ra values:

Process	Ra value of finish
Grinding	0.05–0.8 μm
Finish machining	0.4–1.6 μm
Rough machining	3.2–12.5 μm
Forging/casting	1.6–50 μm

Surface texture comparison plates may be used for assessing the surface texture for most machining processes. A popular set of surface texture plates is illustrated below; it is supplied in an attractive wallet and is for the use of inspectors, draughtsmen and machinists.

A composite set of surface texture plates

Courtesy of Rubert & Co.

EXERCISE 6.9

Select and correctly use surface finish comparison plates to check the surface finish of completed components within 1.6 μm (32 μin).

Trainee to complete

Job title	
Job number	

I have used surface texture measurement tools to measure the following features of my workpieces and have listed my actual measurements below:

Type of surface texture measurement tool	Feature measured	Surface finish required	Measured surface finish

Witness testimony

I confirm that I have seen . . . . correctly select and use the above equipment to measure those features listed by him/her within the tolerances required by this exercise.

Signed . . . . Position . . . . Date . . . .

Slip gauges

Slip gauges are used for instrument checking and a variety of tool setting operations.

They are extremely accurate pieces of hardened steel, ceramic or tungsten manufactured to BS 4311. They must be handled very carefully.

Slip gauges are supplied in boxed sets; a typical workshop set may have 78 blocks.

A 78-piece set of slip gauges

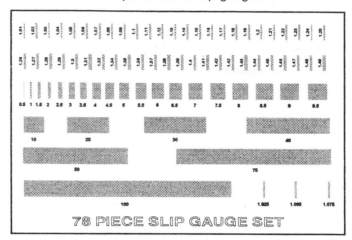

Four quality grades of slip gauges are available:

- calibration
- inspection
- workshop, grade A or B.

Slip gauges can be assembled together making stacks of any size within 0.0025 mm. Once a stack has been used, it is dismantled and the slip gauges are cleaned and lightly oiled and put back into the box ready for reuse. Slip gauges should be stored in a dust free and dry environment at 20 °C.

Some sets have two additional 'wear blocks' usually 2 mm wide. These are put on the top and bottom of the stack to protect the inner slip gauges. It is good practice to use the same face of the wear block as the reference face. Wear blocks are relatively cheap and can be replaced. easily when worn.

Procedure of building up slip gauge stack

Example: (31.285 mm):

(a) Select a slip gauge to eliminate the smallest decimal. — 1.005 mm slip gauge.

(b) Recalculate total. — 31.285 − 1.005 = 30.28 mm

(c) Select a slip gauge to eliminate the next smallest decimal (i.e. the 0.28). — 1.28 mm slip gauge.

(d) Recalculate total. — 30.28 − 1.28 = 29 mm

(e) Select slips to total 29 mm that remains — and 20-mm slip gauges.

(f) Add up the selected slip gauges' heights to check total. — 1.005

1.005
1.28
9.00
20.00
_____
31.285

1.28
9
+20
31.285

Wringing slip gauges

To assemble a stack of slip gauges

(a) Select the required gauges and close the box.
(b) Wipe the slip gauges clean with a soft cloth.
(c) Hold the two widest gauges at 90° to each other, press together and turn as shown – this is called *wringing*. The gauges will stick together.
(d) Continue wringing the slip gauges in decreasing order of size.

Examples of slip gauges in use:

Checking a micrometer's reading by measuring slip gauges

Setting a DTI's dial to zero

Setting a sine bar at a known angle

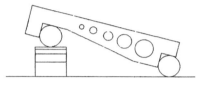

EXERCISE 6.10

Select and correctly use slip gauges to check linear dimensions and parallelism within the tolerances stated on the drawing.

Trainee to complete

Job title	
Job number	

I have used. . . . . to measure the following features of my workpieces and have listed my actual measurements below:

Make/grade of slip gauges	Feature measured	Size and limits	Measured size

Witness testimony

I confirm that I have seen . . . . .correctly select and use the above equipment to measure those features listed by him/her within the tolerances required by this exercise.

Signed . . . . .Position . . . . Date . . . .

A DTI mounted on a cast iron base

Courtesy of Starrett

Dial test indicator

A dial test indicator (also called *DTI*, *Dial Gauge* or *Clock*) is a precision and delicate instrument that has a plunger projecting out of the bottom and a dial commonly graduated in increments of 0.01 mm (or 0.0005″). When the plunger is moved, a rack and pinion mechanism causes the pointer to rotate around the dial, showing the actual amount of plunger movement. A smaller pointer registers the number of whole rotations the pointer makes (usually 1 mm of plunger movement per revolution of the large pointer). DTIs must be held firmly in a static position and a magnetic base is often used for mounting DTIs as it allows flexibility in positioning the instrument. When used for checking components in a production situation, a cast iron clock stand would be used.

The rack and pinion mechanism inside a DTI

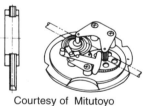

Courtesy of Mitutoyo

Standard DTI

Courtesy of Mitutoyo

DTIs are used in a number of ways

Checking roundness using a DTI and a vee block

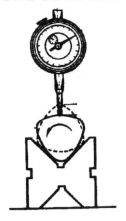

- For checking if a component is within its prescribed limits.
- For checking roundness using a Vee block as shown.
- For setting work level on an angle plate.
- For setting machine vices (see Machining by Milling: Checking runout with a DTI page 322)
- For accurately setting work in a lathe 4-jaw chuck (see Machining by Turning page 312).
- For checking for run out.

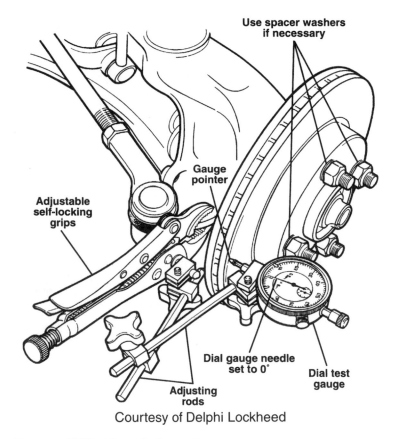

Courtesy of Delphi Lockheed

To use a DTI with a clock stand

- Assemble the slip gauges and stand them on the base under the DTI.
- Lower the DTI until its stylus touches the slip gauge's top surface.
- Carefully lower the DTI further until the pointer goes no more than half a turn round the dial.
- Turn the outer bezel and set it to read zero (i.e. the 0 mark lines up with the pointer).
- Remove and replace the slip gauges to check that the DTI still reads zero.
- Remove the slip gauges and insert the workpiece under the DTI's stylus and note the reading on the dial.
- Any deviation from the slip gauges' size will be indicated on the dial, accounting for the drawing's permissible limits and you will be able to decide if the work is:

- good (i.e. within limits)
- scrap (i.e. too small)
- for rework (i.e. too big).

Other types of DTIs

Now available are DTIs with digital read-out, which are more expensive but easier to read and reset. They are capable of displaying the plunger's position in inches or millimetre units and have a self-contained battery.

A digital indicator

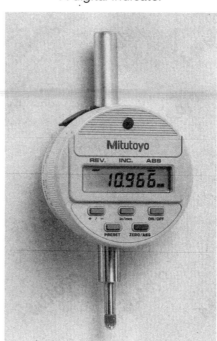

Courtesy of Mitutoyo

DTIs are made to enable a multitude of comparison tasks to be undertaken. Below are some of the more unusual DTIs that are available.

A finger or lever-type DTI

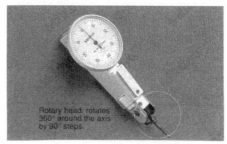

Courtesy of Mitutoyo

A range of special purpose DTIs

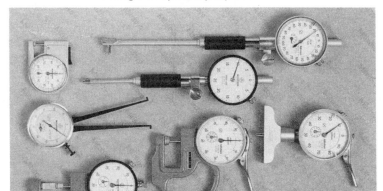

EXERCISE 6.11

Select and correctly use DTI (external, internal) to check linear dimensions and parallelism and concentricity within ± 0.025 mm (or 0.001″) and to set components square or level in a vice.

Trainee to complete

Job title	
Job number	

I have used various types of DTIs to measure the following features of my workpieces and have listed my actual measurements below:

Type of DTI	Feature measured	Size and limits	Measured size

State the advantage of using electronic measuring instruments.

Witness testimony

I confirm that I have seen. . . . . correctly select and use the above equipment to measure those features listed by him/her within the tolerances required by this exercise.

Signed . . . . Position . . . . Date . . . .

7 Marking out for engineering activities

A draughtsman produces detailed drawings of components on paper. The engineering craftsman then proceeds with the manufacture of these components. Marking out is the process of marking guidelines onto the surface of the component material, called the *workpiece*, before it is cut. These lines indicate the overall shape of the finished workpiece and the position of hole centres. Craftsmen carry out marking out before starting most jobs.

The first stage in marking out is to gather all the information about the work you are going to carry out. Normally, you need an engineering drawing and any other relevant work instructions; look to see if there is a safety warning or special instructions. You then need to extract the position of the datums from the drawing.

Datums are the reference positions from which all measurements in each plane are taken. All the component's features should be marked out with reference to their datums. Drawings should identify the component's datums – usually height, length and width. For rectangular workpieces, the datums for height and length are usually two straight faces that are square (at 90°) to each other. For cylindrical and symmetrical work, one datum may be its centre line. Features on the end of a circular component frequently have a point as their datum.

Look at the drawings below and note that all the dimensions in each direction originate from a face, a line or a point.

Face datum (or edge datum)

E.g. rectangular workpiece. All lengths and heights are dimensioned from two faces

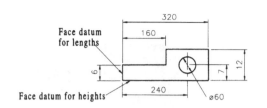

Line datum

E.g. symmetrical workpiece. Height dimensioned from the workpiece's centre line. Lengths dimensioned from the hole's centre line

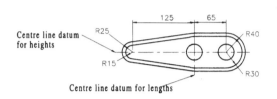

Point datum

E.g. circular workpiece. Features on the end face are dimensioned from its centre point

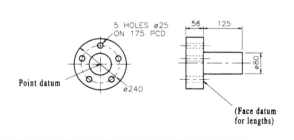

You then need to acquire the workpiece. When you are issued with your workpiece, clean it up and check it to make sure there are no obvious defects visible, you could save a lot of time here!

When marking out workpieces, the same datum faces, lines or points are normally used as those on the drawing. The preparation of the datums on the workpiece is the initial stage in marking out. It must be carried out accurately as all subsequent lines will be positioned relative to the datums. First, remove any protective coating and any dirt, rust or grease. If you are using two face datums, they are usually prepared by filing them flat and square (at 90°) to each other. Any burrs and sharp edges must be removed for safety and accuracy. When using a line or a point datum, the datums are initially marked out onto the workpiece.

Marking out of all other lines can only proceed once the datums have been carefully and correctly established. The most frequently used marking out tools are described in this chapter, together with the methods to use them.

Marking out tools

For accuracy, all marking out tools should be of a high quality and well looked after (i.e. be cleaned and returned to their case or tool store after use). You should always check your tools for condition and serviceability before use (e.g. checking the point on a scriber before use).

It is common practice to coat surfaces being marked out with a quick drying **blue dye** or **copper sulphate**, which makes the scribed lines clearer and easier to see. The coating may be applied with either a brush or by aerosol spray.

Pens and pencils

Pencils and pens are not normally used by engineers for marking out as they are not accurate enough. However, they do not scratch the surface of the work and they leave a removable impression. For these reasons, they are occasionally used to mark out materials in which scratching is **not** desirable, for example when marking out:

- plated steel, to avoid possible corrosion arising from scratching through the protective coating
- thin or fragile materials, which could be damaged by the scriber's point
- sheet metal that is to be bent; a scored line made by the scriber's point may cause the material to crack, which can be avoided by using a pencil.

A typical engineer's scriber

Scribers

Scribers are used by engineers together with guiding instruments such as rules, try squares, protractors and radius gauges for scribing (scratching) fine permanent lines onto the surface of the work. Scribers have a finely ground, hardened and tempered point that should be kept sharp; they are made from high carbon steel and are about 120-mm long. They are held in the same way as pens and have a knurled shank for grip.

When scribing lines, the scriber's point should fit between the guiding instrument and the work.

Safety Note: Never carry scribers in your pockets. Carry them with the point facing down.

Rules

Made from steel, an engineer's rule is a simple and easy tool to use for guiding a scriber when marking out straight lines. A rule can be used to assist marking out to within ±0.5 mm if care is taken. When marking out length with a rule and a scriber, ensure that the rule is held at 90° to the work's datum face. The end of the rule is used to guide the scriber as shown in Figure a.

a. Correct *method of marking length of 44 mm with a rule.*

b. Incorrect *method of marking out length of 44 mm with a rule; the scribed line will not be straight or square.*

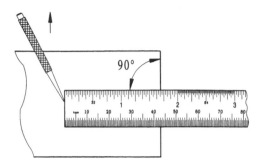

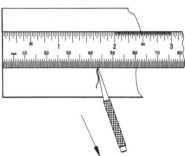

Care for your rule! The sides and end of all rules should be protected from wear and not used for turning, prising or cleaning anything that would damage the rule.

Tape rules are used for marking lengths on materials for which an engineer's rule is too short, they cannot be relied on for scribing straight lines as they are not rigid.

Using a scriber with a try square

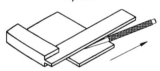

Engineer's squares

There are three types of engineer's squares in use for marking out: try squares, centre squares and box squares. The try square is described in 'Checking workpieces for accuracy'; the use of a centre square and box square for marking out is described here.

A combination set fitted with a centre head

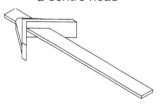

Centre square

Centre squares are used to mark out the centre points at the end of cylindrical bars. They can be supplied as part of a combination set or as specialist instruments.

The procedure to mark out the centre point of the end of a cylindrical bar using a centre square is as follows:

(a) Locate centre square flat on end of bar, vee faces resting on bar's curved edge.

(b) Scribe line along straight edge (centre line of bar).

(c) Rotate work 90°, scribe second centre line.

(d) Centre point at intersection of centre lines.

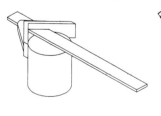

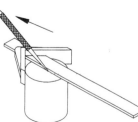

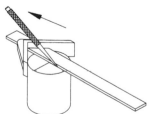

Box square

A box square is a tool used to mark out lines on the side of a cylindrical workpiece parallel to the cylinder's axis. Box squares are suitable for marking out keyways and slots on round bar and tube.

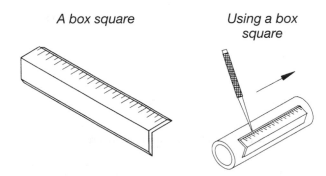

A box square

Using a box square

EXERCISE 7.1

Follow the pre-planned sequence of operations to mark out the centres of the three holes on the drilled bar shown below in Drawing No. Mo1.

Your lines should be clearly marked out.

Dot punch the outline at intervals of about 15 to 20 mm and at each intersection.

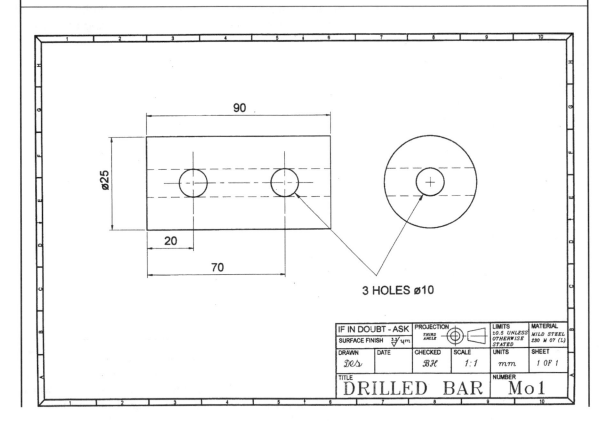

90

ø25

20

70

3 HOLES ø10

IF IN DOUBT - ASK	PROJECTION *THIRD ANGLE*		LIMITS ±0.5 UNLESS OTHERWISE STATED	MATERIAL *MILD STEEL* 230 M 07 (L)
SURFACE FINISH ³·³√μm				
DRAWN *DES*	DATE	CHECKED *BH*	SCALE *1:1*	UNITS *mm*	SHEET *1 OF 1*

TITLE	NUMBER
DRILLED BAR	Mo1

Planned sequence of operations:

No.	Task/line	Tools used
1.	*Remove all burrs and sharp edges from the component.*	*Files & emery cloth.*
2.	*Gather all tools and check them over for safety, accuracy and suitability.*	*Squares; rules; files & scribers.*
3.	*Prepare datum face.*	*File, try square.*
4.	*Mark out centre line on side of bar.*	*Box square.*
5.	*Mark out centre line on end of bar.*	*Centre square.*
6.	*Mark out hole centres from bar's end.*	*Rule; scriber.*
7.	*Mark out centre of hole on end of bar.*	*Centre square.*
8.	*Check finished work.*	*Rule.*

Witness testimony

I confirm that the drilled bar has been marked out accurately within the stated limits and that. . . . . .removed all waste and surplus materials, leaving the work area in a safe and tidy condition.

Signed. . . . . .Job title . . . . Date . . . .

Drawing an external radius with a scriber and a radius gauge

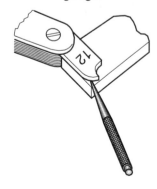

Radius gauges

To mark out an arc of a specific radius, the profile of a radius gauge can be scribed around.

Care must be taken to ensure that the point of the scriber follows the profile of the gauge. The procedure is not entirely satisfactory as it can result in limited precision. A preferred method of scribing arcs is to use dividers.

Odd leg calipers

Odd leg calipers are also known as *Hermaphrodite* or *Jenny calipers* and are used for scribing lines parallel to the datum face of a work-piece. The scribing leg of odd leg calipers has a sharp point, which is sometimes replaceable. The other leg is shaped in the form of a projecting heel or hook for locating on the work's datum face.

Setting odd leg calipers on a rule

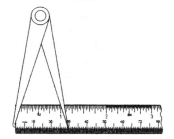

To set odd leg calipers, the distance from the heel to the point is set with a rule. It should be noted that the engraved lines on the rule enable the point of the odd legs to click into position to ensure an accurate setting.

When correctly set, the odd leg caliper is held at 90° to the work, its heel against the datum face of the work and its point on the work. The odd leg caliper is moved along the face of the work so that a single straight line is scribed parallel to the work's face (see figure). Set with a rule, odd leg calipers can be used to mark out lines accurately within ±0.5 mm.

Odd leg calipers used to scribe a line parallel to a datum face. Note the two different heel types

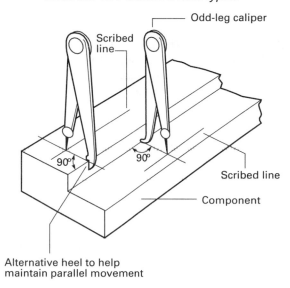

Using odd leg calipers to mark out hole and radius centres

When marking out a hole or a radius, its centre point must first be located. The centre point is the point at which the hole's or radius' horizontal and vertical centre lines intersect. The component in the figure shows holes and a radius. The procedure for marking out their centre points is shown below:

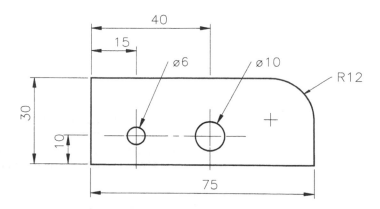

1. To mark out centre points of holes:

(a) Prepare datum faces A and B by filing straight and square.

(b) Set odd leg calipers at 10 mm (distance from face A to both holes' centre line).

(c) Scribe hole's centre line from face A.

(d) Set odd leg calipers at 15 mm (distance from face B to ø6 hole's centre line).

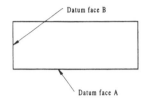

(e) Scribe ø6 hole's centre line from face B.

(f) Set odd leg calipers 40 mm (distance from face B to ø10 hole's centre line).

(g) Scribe ø10 hole's centre line from face B.

(h) Intersections of centre lines are hole's centre points.

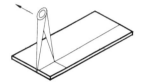

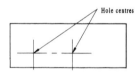

2. To mark out centre points of radius:

(a) Note the size of the radius on the drawing (12 mm).

(b) Calculate distance from radius' centre to face B.

$$75 - 12 = 63\,mm$$

(c) Set odd leg calipers at 63 mm. Scribe the radius' centre line from face B.

(d) Find distance from radius' centre to face A.

$$30 - 12 = 18\,mm$$

(e) Set odd leg caliper at 18 mm. Scribe centre line.

(f) Point at which centre lines intersect is radius' centre.

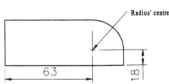

After the centre point has been accurately located, it can be punched and used for guiding a drill or for further marking out procedures.

Dot punch and centre punch

Dot and centre punches are used to punch dimples during marking out. They are held vertically and lightly hit with a $\frac{1}{2}$ lb (125 g) hammer.

A dot punch in use punching a dimple at a centre point

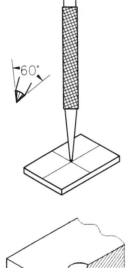

Each type of punch produces a different type of dimple, but they are both made of hardened and tempered 0.8% carbon steel, about 100-mm long and knurled to give a positive finger grip.

The **dot punch** has a 60° point and is used to:

- punch dimples at the centre point of holes and radii for locating one of dividers' points when scribing circles and radii
- punch small, equally spaced dimples along scribed lines at frequent intervals to enable the marked lines to be identified during cutting operations. Once cut, the remaining half of the dimple acts as a witness mark to show that the cutting operation has been done accurately.

Dot punches are also used to punch dimples for **marking out** lengths with dividers.

The **centre punch** has a 90° point. It is used **after** dot punch to increase the diameter of the dot-punched dimple. The larger dimple is to locate the point of a drill in position at the start of a drilling operation.

Centre punches are used as follows to mark out the hole's centre point:

- Dot-punch a **small** dimple at centre point.
- Check position of dot-punched dimple.
- Repunch the dot-punched dimple with centre punch.

A centre punch

Section through a centre-punched hole

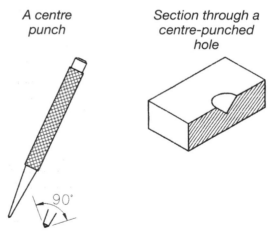

A pair of engineer's dividers

Dividers

Dividers are used to mark out circles, arcs and repeated pitches by scribing arcs. They are set within an accuracy of ±0.5 mm with an engineer's rule. Each of the dividers' legs has a sharp point, and the instrument is held and turned by a peg on the bow.

When marking out a circle with dividers, the procedure is as follows:

(a) Mark out and dot-punch the circle's centre point.

(b) Set the distance between the dividers' points to the circle's radius with a rule (the dividers' points should click into one of the rule's engraved lines).

(c) Locate one point of the dividers' legs into the dot-punched dimple.

(d) Rotate the dividers by the peg to scribe a true circle.

Using dividers to mark out a circle

- Dot-punch circle's centre point.
- Set dividers at circle's radius.
- Locate one point in a dot-punched dimple.
- Scribe the circle.

EXERCISE 7.2   Marking out the drill drift

Follow the pre-planned sequence of operations to mark out the drill drift shown below in Drawing No. Mo2.

Your lines should be clearly marked out within ±0.5 mm.

Dot punch the outline at intervals of about 15 to 20 mm and at each intersection.

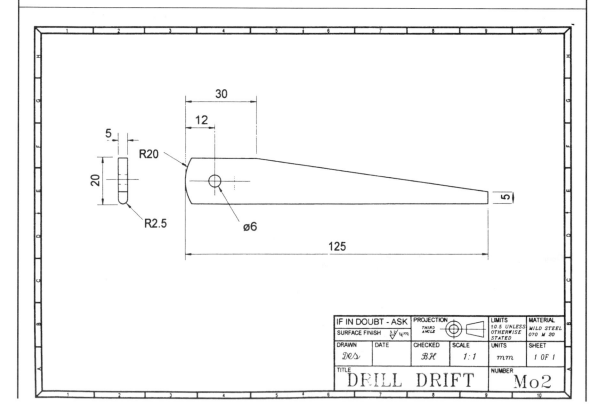

IF IN DOUBT - ASK	PROJECTION THIRD ANGLE		LIMITS ±0.5 UNLESS OTHERWISE STATED	MATERIAL MILD STEEL 070 M 20
SURFACE FINISH				
DRAWN *DES*	DATE	CHECKED *BH*	SCALE *1:1*	UNITS *mm*	SHEET *1 OF 1*
TITLE DRILL DRIFT			NUMBER Mo2	

Pre-planned sequence of operations for marking out the drill drift:
(This plan is completed for your guidance. In future exercises, you must plan the sequence of operations by completing similar tables **before** starting the marking out.)

No.	Task/line	Tools selected
1.	Remove all burrs and sharp edges from the component.	File.
2.	Gather all tools and check them over for safety, accuracy and suitability.	
3.	Prepare datum faces.	File, try square.
4.	Mark out centre line.	Odd leg calipers & rule.
5.	Mark out 20-mm radius' centre.	Odd leg calipers & rule.
6.	Mark out ø6-mm hole centre.	Odd leg calipers & rule.
7.	Mark out overall length.	Rule, scriber & try square.
8.	Mark out angled edge.	Rule & scriber.
9.	Dot and centre punch centre points of hole and radius.	Dot punch & $\frac{1}{2}$ lb. hammer.
10.	Mark out radius and hole outline.	Dividers & rule.
11.	Make all outlines more clear.	Dot punch & $\frac{1}{2}$ lb. hammer.
12.	Centre-punch hole centre.	Centre punch & $\frac{1}{2}$ lb. hammer.
13.	Remove waste and surplus materials.	
14.	Ensure that the work area is left in a safe and tidy condition.	

Witness testimony

I confirm that the drill drift has been marked out accurately within the stated limits and that. . . . . .observed specific safety precautions and left the work area in a safe and tidy condition.

Signed . . . . .Job title . . . . Date . . . .

Marking out equally spaced features

When a number of similar features are equally spaced along a line or around a circle, the distance between the features is called the *pitch*.

Linear pitches (straight line pitches)

Example: To mark out six holes, which are equally spaced on a 60-mm line, the procedure would be as follows:

(a) Scribe centre line of holes with odd leg calipers.	(b) Find pitch of features by dividing overall length by number of **spaces.** *Pitch = 60 mm ÷ 7 spaces = 8.57 mm*	(c) Set odd leg calipers at pitch, scribe first hole position.
(d) Set dividers as accurately as possible to calculated pitch.	(e) Dot-punch dimple at hole's centre point.	(f) Locate dividers' leg in dot-punched dimple, scribe arc to cross centre line.
(g) Repeat stages (d) and (e) until all hole positions are marked out.	(h) Check positions of dot-punched dimples, centre punch each centre point.	(i) Holes now ready for drilling.

Pitches on a pitch circle diameter (PCD)

When features are equally spaced on a circle, the circle is called a *PCD*.

Example: To mark out eight equally spaced holes ready for drilling on a ⌀60 mm PCD, the procedure would be as follows:

(a) Mark out and dot punch centre point of PCD (point datum). 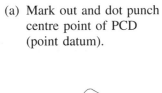	(b) Set dividers at PCD's radius (30 mm) and scribe PCD.	(c) Dot punch a dimple in first hole's position, that is intersection of the PCD and the centre line. 
(d) Calculate pitch of holes by referring to Appendix IV. *Pitch =* *0.3287 × 60 =* *19.72 mm* 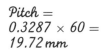	(e) Set dividers as accurately as possible to calculated pitch. 	(f) Locate a divider leg in dimple, scribe arc across PCD. Dot punch this intersection. 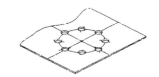
(g) Repeat stage (f) until all hole centres are marked out. 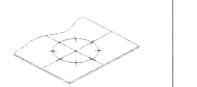	(h) Check position of dot-punched dimples, centre punch each one. 	(i) Holes now ready for drilling.

EXERCISE 7.3

Plan the marking out of the pitches exercise shown on Drawing No. Mo3 using the planning sheet below.

When your plan has been approved, complete the exercise by marking it out within ±0.5 mm. Your lines should be clearly marked out within ±0.5 mm. Dot punch the centres at each intersection.

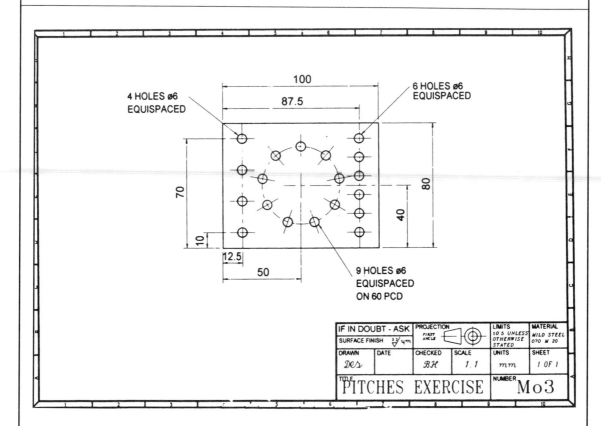

Sequence of operations planning sheet:

No.	Task/line	Tools used
1.		
2.		
3.		
4.		

5.		
6.		
7.		
8.		

Witness testimony

I confirm that the pitches exercise has been planned correctly and marked out accurately within the stated limits, and that.............................observed specific safety precautions and is aware of what can go wrong with marking out equipment and the action to take when things do go wrong.

Signed . . . . Job title . . . . Date . . . .

Surface plate

View of a cast iron surface plate with its protective cover removed

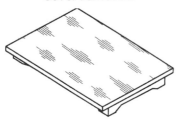

A surface plate provides a large flat reference surface for mounting work and tools during marking out. The flat surface provides stability and enables the marking out tools described in the following pages to be used efficiently. For easy access, surface plates are usually located on or near the workbench.

Surface plates are usually made of cast iron, owing to its self-lubricating and hard-wearing properties. Cast iron surface plates have ribs underneath, which increase their rigidity (see drawing below).

In order to retain its flatness, a surface plate must be treated with extreme care:

View of strengthening ribs underneath a cast iron surface plate

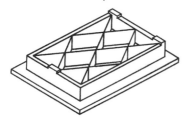

- Always clean a surface plate with a soft dry cloth before use and cover it after use.
- Slide heavy objects onto a surface plate rather than lowering them.
- Do not drop anything onto a surface plate.
- **Never** hammer anything on a surface plate.

Surface plates can also be made from granite or glass. These materials are sometimes preferred as they are non-magnetic, rust free and do not burr. All good-quality surface plates are accurately finished and conform to BS 817 (1988).

Angle plate

An angle plate is used to support a workpiece vertically on a surface plate so that it can be marked out accurately. The work can be clamped to an angle plate when marking out from a line or point datum or they can be used for resting workpieces against when marking out from a face datum.

Angle plates are made from cast iron in various designs. They are usually drilled or slotted so that bolts can go through them for clamping the workpiece. The surfaces of angle plates are accurately machined so that they are square (at 90°) to one another.

Three designs of cast iron angle plates

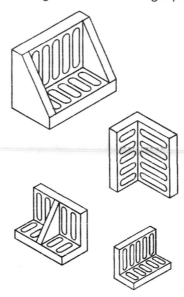

Work set up for marking out using an angle plate on a surface plate

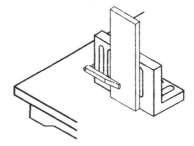

Scribing blocks and surface gauges

A scribing block is used on a surface plate for marking out parallel lines on workpieces. Scribing blocks are made with a cast iron or hardened steel base and have a vertical steel spindle. A special double ended scriber is clamped to the spindle in such a way as to enable the height of the scriber's point to be set.

A surface gauge and a scribing block (inset)

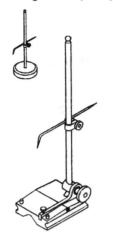

A scribing block is known as a *surface gauge* if it has a means of **fine** adjustment for setting the height of the scriber's point.

When marking out, the work and rule are set vertically against the angle plate. The height of the scriber's point is adjusted as required. The instrument is then moved along the surface plate, scribing a line on the workpiece as shown in the figure.

The usual method of setting a surface gauge for marking out

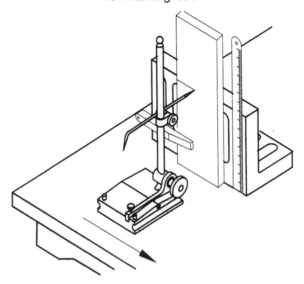

Safety note! To avoid possible eye injury when using a surface gauge, wear safety glasses.

The unused scribe point can be covered with a cork or rubber.

Some common methods of setting and using surface gauges

- Scribing lines parallel to a surface.

- Marking out a casting.

- Setting to a combination square rule.

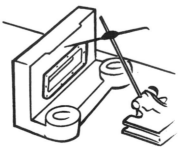

- Using the setting pins to scribe parallel to an edge.

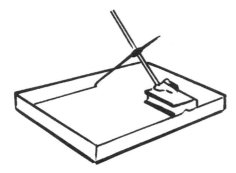

- Checking a surface for parallelism.

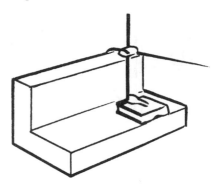

Parallels used to support

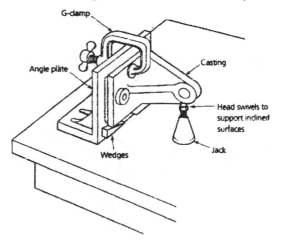

Auxiliary tools for setting irregular-shaped workpieces

- **Parallel bars** – Parallel bars are used for supporting workpieces when the workpiece's shape makes it difficult to mount on a surface plate. They are made in matching pairs from hardened steel, and are accurately ground to size and stored and maintained as a pair. A workshop would normally have a range of sizes of parallels available, enabling all types of workpieces to be set up.
- **Jacks and wedges** – These devices can be used for supporting heavy irregular-shaped workpieces as shown in the diagram. They are adjustable for height within a range and help when setting component's level within fine limits.
- **Dial test indicator** (DTI) – A DTI can be used to compare the position of each end of a component that is being set up on an angle plate to ensure that the work is set level. DTIs are also used for setting work straight and level in milling machine vices. (See page 322).

Jacks and wedges used to locate a workpiece

G-clamp

Angle plate

Casting

Head swivels to support inclined surfaces

Wedges

Jack

EXERCISE 7.4   Stepped block

Follow the pre-planned sequence of operations to mark out the stepped block shown in Drawing No. Mo4.

Your lines should be clearly marked out within ±0.5 mm.

Dot punch the outline at intervals of about 15 to 20 mm and at each intersection.

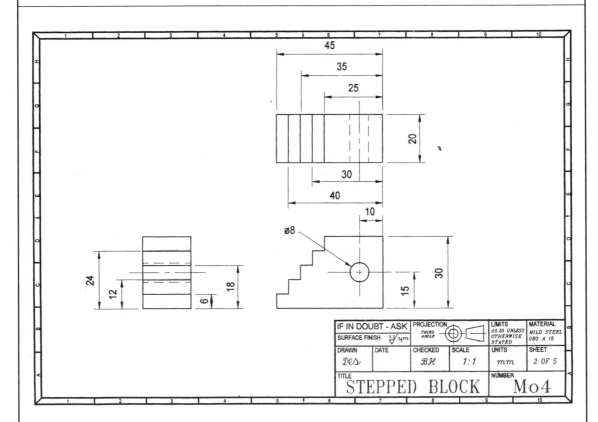

Planned sequence of operations:

No.	Task/line	Tools used
1.		
2.		
3.		
4.		

5.		
6.		
7.		
8.		

Witness testimony

I confirm that the stepped block has been marked out accurately within the stated limits and that . . . . removed all waste and surplus materials and observed the specific safety precautions to be taken when using marking out mediums.

Signed . . . . Job title . . . . Date . . . .

The vernier height gauge

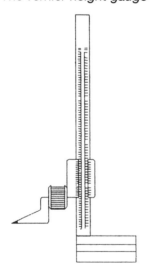

Vernier height gauge

Vernier height gauges are used to mark out lines when accuracy is particularly important. Vernier height gauges have a large steel base for stability and a hardened steel column engraved with inch and mm units. The moving slide is engraved with vernier calibrations enabling settings to an accuracy of 0.02 mm and 0.001″. Attached to the moving slide is a wedge-shaped steel scribe that has a pointed tungsten carbide tip. The height of the scribe's point can be finely adjusted with a thumbscrew.

The method of setting and reading a vernier height gauge is the same as for other vernier instruments, which are described in detail in the checking workpieces for accuracy chapter, pages 160–62.

The work to be marked out with a vernier height gauge must be set vertically on a surface plate, using an angle plate.

The height of the scribe is set with the moving slide's fine adjusting screw. The marking out lines are then scribed onto the work's surface by steadily moving the vernier height gauge along the surface plate (following the same procedure as for using a surface gauge).

On some vernier height gauges, the height of the **main** scale can also be finely adjusted. This enables measurements to start from a convenient reading, which is useful if the shape of the workpiece is such that it needs to be mounted on a parallel strip as shown below.

Note that digital height gauges are now becoming more common as they are much quicker to set up and more accurate to set.

Marking out work using a parallel bar

Set vernier height gauge at known
height by touching on top of parallel bar
and setting to a convenient reading

Mount work onto parallel bar, mark out
all horizontal lines (remember to add
height of parallel to all settings)

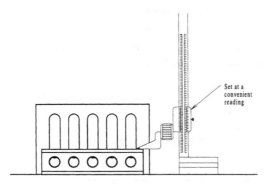

Set at a
convenient
reading

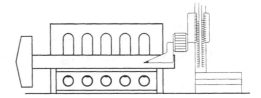

Precision vernier height gauges must be stored in their original packaging case and should be carefully used and maintained. If you suspect that the instrument is inaccurate, it must be reported to your supervisor at the earliest opportunity to reduce any wastage that would otherwise be caused. Remember that the inaccuracy could be your fault, so a 'back footed' approach could prove to be appropriate.

Marking out circles with witness lines

When marking out large holes prior to drilling or boring operations, additional lines can be marked out to show the machinist if his drill is running 'true' or if it is slightly off centre. This involves extra marking out lines called *witness lines*; this procedure is quick and reliable. The following method of marking out witness lines should be undertaken when marking out large holes that are to be drilled, ensuring that the holes are accurately positioned:

Method of marking witness lines

(a) Mark out circle centre lines.

(b) Dot punch intersection of centre lines.

(c) Scribe required circle with dividers.

(d) Scribe a second, slightly larger circle.

The second circle is a witness line. It can still be seen after the hole has been machined. The witness lines are used as evidence to show if the hole is accurately positioned.

Drill running true	Drill running below centre	Drill running to the left

EXERCISE 7.5   Marking out the depth gauge body

Follow the pre-planned sequence of operations to mark out the depth gauge body shown on Drawing No. Mo5.

Your lines should be clearly marked out within the drawing limits.

Dot-punch the profile at intervals of about 5 to 10 mm and at each intersection.

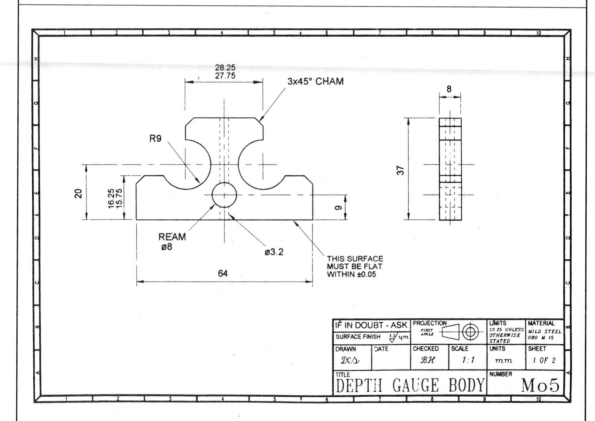

Planned sequence of operations:

No.	Task/line	Tools used
1.		
2.		

3.		
4.		
5.		
6.		
7.		
8.		

Witness testimony

I confirm that the depth gauge body has been marked out accurately within the stated limits and that . . . . removed all waste and surplus materials and left the work area in a safe and tidy condition.

Signed . . . . Job title . . . . Date . . . .

Workshop protractor

Workshop protractors can be used for marking out angles on workpieces. The reading and setting of workshop protractors is described in detail in Checking completed workpieces for accuracy. When marking out with a workshop protractor, an accuracy of about ±0.5° (called 30 minutes) can be readily achieved. A workshop protractor can be used as:

- a hand-held tool, to guide a scriber
- a work setting device on a surface plate

Using a workshop protractor as a hand-held tool

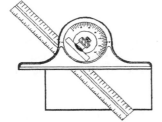

Using a workshop protractor as a work setting device

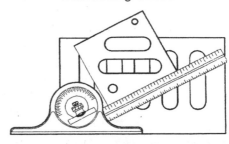

Vee blocks

Vee blocks can be used for mounting cylindrical work (shafts) on surface plates so that the work can be accurately marked out on its sides or ends. They are also used to support rectangular work by one corner so that the work's sides are held at 45°. This enables lines to be accurately scribed on the work's surface at 45° with a surface gauge, scribing block or vernier height gauge.

A pair of vee blocks and their clamp

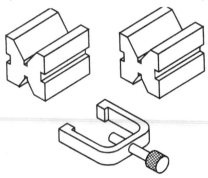

Vee blocks have their surfaces accurately ground square to one another. There is a vee-shaped groove in at least one side into which the work is located. Made in either cast iron or case-hardened mild steel, vee blocks are usually supplied in matched pairs, often with a clamp to enable the work to be secured in position.

Work mounted longitudinally on a vee block

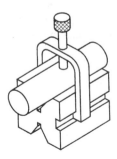

Work mounted vertically on a vee block

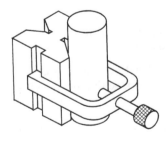

Vee block used to support work with square corners at 45°

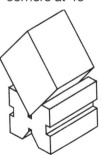

Vee blocks can be used to mount shafts or cylindrical work for marking out and drilling. The shaft is mounted vertically for drilling along its axis or mounted horizontally for drilling across its axis.

EXERCISE 7.6 Marking out the angular slotted plate

Follow the pre-planned sequence of operations to mark out the angular slotted plate shown in Drawing No. Mo6.
Your lines should be clearly marked out within ±0.5 mm.
Dot-punch the outline at intervals of about 15 to 20 mm and at each intersection.

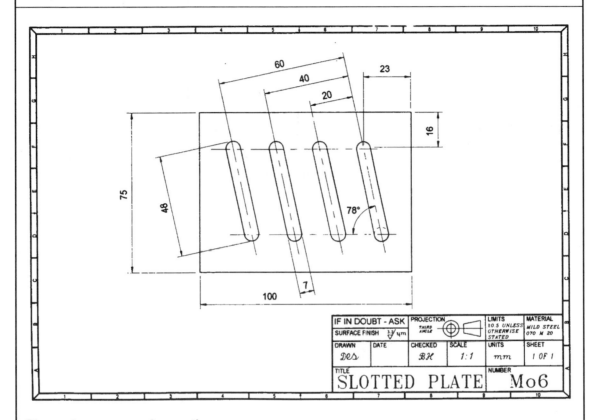

Planned sequence of operations:

No.	Task/line	Tools used
1.		
2.		
3.		
4.		

5.		
6.		
7.		
8.		

Witness testimony

I confirm that the slotted plate has been marked out accurately within the stated limits and that . . . . removed all waste and surplus materials and left the work area in a safe and tidy condition.

Signed . . . . Job title . . . . Date . . . .

8 Fitting using hand skills

Exercise checklist

(Ask your assessor to initial the lower box for each exercise when you have completed it.)

Exercise no.	8.1	8.2	8.3	8.4	8.5	8.6	8.7	8.8	8.9
Initials									
Date									

The bench fitter is one of the most skilled tradesmen in mechanical engineering. Bench fitting involves using hand tools for hand-shaping and fine-finishing of components, so that they can be assembled with precision. Typical components assembled by fitters include machine parts, measuring devices and work-setting fixtures. The fitters' work can also involve hand-making individual parts as required. Fitters use simple tools such as files, saws, hammers and punches; they also use light machine tools such as drilling machines and shapers. The bench fitter is based at a workbench fitted with a vice and usually has access to a surface plate for marking out workpieces.

The hand skills used in fitting are not easy to acquire and must be practised so that a high level of precision can be achieved. Knowledge of the tools and equipment is necessary so that the right tool can be selected for the job in hand. The following pages describe various hand tools and their uses. There are a series of exercises designed to enable you to demonstrate your skills in selecting and using these tools.

Bench and vice

Benches are work tables used by fitters that must be steady and strong. Most benches are made of thick wood and are approximately 1 m high. The top surface of the bench must be able to withstand shock because it may be subjected to sudden impact when work is being hammered or chiselled. Many benches have a vertical tool rack and a drawer underneath for storage of tools and workpieces.

A fitter's bench would normally have a vice fitted. The vice is used to hold the work steady when it is being shaped. The vice must be securely fitted to the bench far enough forward to enable the work to extend below the top surface of the bench as shown. For maximum efficiency, the vice's height should be the same as the fitter's elbow when he is working.

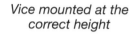

Vice mounted at the correct height

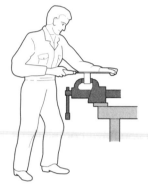

Standard serrated jaws

Protective fibre grip inserts

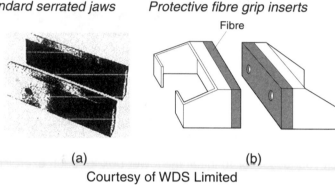

(a)   (b)

Courtesy of WDS Limited

There are various designs of vices – most are of cast iron and fitted with hardened steel jaws. These jaws may have a plain smooth surface or be engraved with serrated grips. If the vice holding the workpiece has serrated jaws, a pair of inserts may be used to protect the work that may be made from a soft material, for example aluminium or fibre.

Cutaway view of a record vice

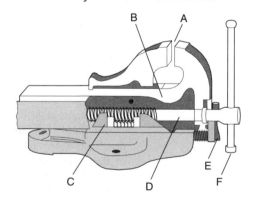

Features (cutaway view)

A Hardened steel jaws
B Precision moving jaws
C Split nut for fast opening and closing
D Steel screw with buttress type thread
E Quick release lever
F Steel handle

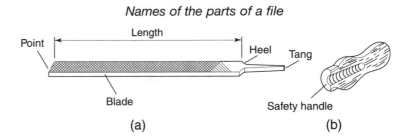

Names of the parts of a file

(a) (b)

Files and filing

A file is a hand tool used to remove material from workpieces. Various types of file are available, and all are made from hardened and tempered high carbon steel. In the hands of a skilled fitter, very small quantities of material can be removed to accurately finish a workpiece to its required shape and size.

File handles

A file must always be fitted with a suitably sized handle fitted to the tang. This is for comfort and safety. Never use a file without a handle because the file's tang can seriously injure the palm of your hand. Files should never be used on a centre lathe.

Most file handles are wooden and have a metal ring around the front called a *ferrule* to stop the handle splitting. A split handle could cause splinters in your hand. Foreign matter embedded in a file's handle can also cut your hand; so always check the handle before starting work.

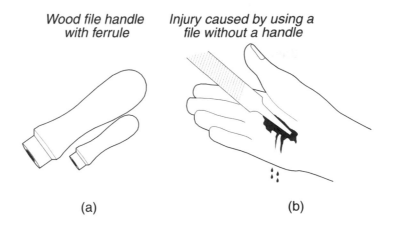

Wood file handle with ferrule *Injury caused by using a file without a handle*

(a) (b)

Selection of file

Most fitters keep a range of files at their work area that are described by their **cut** of the teeth, **shape** of the section and **length** of the file. The correct file for the job in hand should always be selected. The selection is determined by the material being filed and the shape of the finished workpiece.

For example:

A long file would normally be used for large workpieces; shorter files are for smaller intricate work.
A coarse file will remove material more quickly; smooth files are for finishing accurately to size.

File cuts

The pattern of the teeth on the file's body is the cut.

Double cut	**Single cut**	**Rasp**
For most ferrous metals. Most popular file cut and is widely available	*For soft non-ferrous metals. Stays sharp longer as the teeth are long edges, not points*	*For roughing very soft materials, e.g. aluminium or plastic. Leaves a rough finish*
	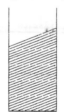	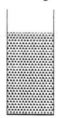

Grade of file cut. Files are graded by the pitch (i.e. the spacing) of their teeth. The three most common grades and their uses are shown below. When gauging one file with another, the two files must be of the same length, because the pitch of the teeth on the same grade of a file varies with its length.

Bastard	**Second cut**	**Smooth**
Heavy metal removal, rough finish	*General purpose, fair finish*	*Fine finishing of work needing a smooth surface finish*

Note. Rough and dead smooth grades are less commonly available.

File shape

The most common **file shapes** and their uses are shown below:

Hand
General filing. One edge has no teeth (safe edge) that prevents a shoulder from being filed or scratched

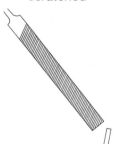

Flat file
General purpose. Being barrelled, it can be used for filing surfaces that are precisely flat

Ward
Filing inside narrow slots and grooves. Narrower than most other files

Square
For rectangular holes and slots with square corners

Swiss (needle) files
These miniature files are used for finishing work and smaller workpieces. A large range of shapes are available

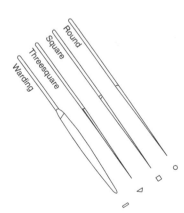

Half round
Similar use as the flat file on front, curved side is used for large internal curves

Round
For small internal curved surfaces and hole enlargement

Three square (triangular)
For finishing sharp corners and angles

There are also other shapes of file produced that are available for special types of work.

Length of file

The length of files is measured as the distance from the tip to the heel of the file (the length measurement of needle files includes the handle). A long file (approximately 12″ or 14″) would be selected for large work. A short file (of say 4″ or 6″) would be selected for intricate work and for accurate finishing.

Using files

When using a file, the workpiece is first secured in a suitable vice. If the vice is set at the correct height, it will be much easier to file a surface flat.

- If the workpiece is set too high in the vice, the fitter will file upwards causing the workpiece to be out of square.
- If the workpiece is set in the vice at the correct height, the fitter can file flat more easily, producing square workpieces.
- If the workpiece is set too low in the vice, the fitter will file downwards causing the workpiece to be out of square.

The file is pushed over the work at an even and comfortable rate, applying downward pressure on the forward stroke and relieving the pressure on the return stroke. Use the full length of the file whenever possible. A number of different techniques are used by fitters as described below:

Straight filing – This involves gripping the file in one hand and using the other hand to guide the file lengthways across the work; the cutting pressure is applied with the palm of the hand. The feet are positioned for comfort and balance. When finishing, apply pressure on the forward (away) stroke lightly with both hands as shown and relieve the pressure on the return stroke as before.

Roughing – heavy
pressure

Finishing – light
pressure

Filing curves and radii is best done from the side of the vice, rocking the file as it is passed over the work. The surface should be compared to a radius gauge at regular intervals to check if the material is being removed from only the correct places.

Checking with a radius gauge

Correct:
radius filed
matches radius
gauge

Radius too large:
too much metal
has been removed

Radius too small:
file with more
material off

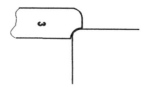

When filing curves and radii, stand in line with the vice jaws and swing the file as shown:

Start with elbow raised high

Midstroke

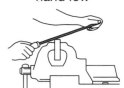

Finish stroke with hand low

Draw-filing – This is a technique used to finish surfaces to a high-quality surface finish. A surface finish of $0.8\,\mu m$ can be achieved by drawfiling with a smooth file. Hold the file firmly at both ends and push and pull the width of the file along the length of the work.

Draw-filing a flat workpiece

Filing flat – To file an accurate flat surface takes much practice and patience. First, smear marking blue onto the top surface of a surface plate and rub the work on the 'blued' area. The marking blue will be transferred to the workpiece at the points of contact known as *high spots*. The workpiece is then mounted in a vice and the high spots are removed with a file – this process is repeated until the blue marks on the workpiece appear evenly spread over the surface being filed flat.

It is good practice to support the workpiece vertically with an angle plate while rubbing the workpiece on the surface plate. When finish filing a broad surface flat, use a flat file because it is slightly barrelled in thickness, allowing the fitter to file parts of the surface without removing the material from other areas.

Workpiece held against an angle plate and rubbed on a surface plate

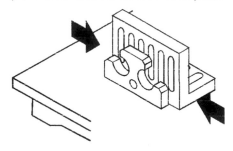

The high spots are now 'shiny'

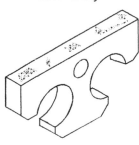

Holding a try square against the work and noting the high spots

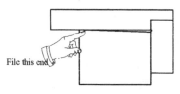

File this end

Filing square – As the name suggests, filing square involves filing the edges of workpieces square to one another, often to form a datum for marking out. Hold the work and a try square up to light and view the gap between the work and the try square's blade. Notice where the light passes through and where the workpiece makes contact with the try square's blade. Grip the workpiece in a vice and lightly file away the high spots (areas of contact) so that the whole length of the try square's blade and stock are in contact with the workpiece.

Measuring the finished size with a micrometer

Finishing – The trainee fitter must take care to only remove material from the areas that are necessary. This seems obvious but is, in practice, a difficult process. The marked lines should be constantly observed and not cut across until the component has been measured. When finish filing a surface's profile, check it frequently against a try square or surface plate and only remove small quantities of material from the 'high spots' at a time.

Care of files

Removing clogged material from a file with a file card

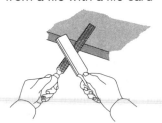

- Store files separately; if files' teeth rub together, they can become damaged. Many fitters hang their files in a rack.
- Keep files dry to stop them from becoming rusty.
- Remove any clogged material from the teeth of files with a 'file card'.
- When working on soft metals, apply chalk to the teeth of files to reduce the chances of clogging.
- Never hit anything with a file as the file could break.

EXERCISE 8.1   Drill drift

Make the drill drift using the procedure and tools indicated. Your finished work should be within the drawing limits. (This workpiece was marked out in Exercise 7.2.)

On completion of the task, carefully measure your work and write in the table the actual sizes, noting any errors.

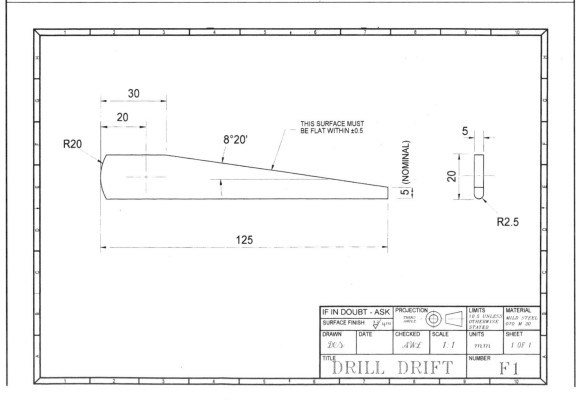

Planning:

Procedure	Tools and equipment selected
1. Acquire material and check for size. 2. Mark out all centre lines and edges. 3. Dot punch the profile. 4. Check the profile. 5. Saw cut the rough profile. 6. File to clear off marked lines. 7. Finish file profile and check against drawing. 8. Stamp name, deburr and polish. 9. Clean the work area.	*Marking out tools & equipment:* • *Metric 300-mm rule* • *Odd leg caliper* • *Scriber.* • *Dot punch.* • *1/2 lb. hammer.* *Range of files.* *Radius gauges.* *Letter stamps*

Complete the tables with the correct information:

Files used (state cut, shape, length):	*Reason for selection:*
1.	
2.	
3.	
4.	
Height of the top of the vice from floor:	

List here the checks that were made on your tools before starting work to ensure that they are safe and sharp:

1.

2.

3.

List here the waste material disposal procedures that are to be followed in your workshop:

1.

2.

3.

Sketch:	*Hand file*	*Flat file*
Make a dimensioned sketch of a hand file and a flat file showing clearly which type has a 'safe edge'.		

Finished inspection report:

Component dimensions (mm)	*Limits:*	*Actual size:*	*Error:*
1. Overall length 125	*125.5* *124.5*		
2. Overall height 20	*20.5* *19.5*		
4. 20 Radius	*To fit radius gauge*		
5. Blade flatness:	*'Feeler tight' within 0.5 mm over whole length*		

Results:

Is the workpiece to the drawing specification? | YES | | NO |

If completed workpiece is below drawing specification, list below here the reasons for any errors, stating clearly how these errors will be avoided in future.

Start time and date:	End time and date:	Time taken (hours):

Witness testimony

I confirm that the drill drift was made within the stated limits and that. . . . . .cleared the work area on completion of the task.

Signed . . . . .Job title . . . . Date . . . .

Hacksaws

A hacksaw is a handsaw that is effectively used when large amounts of material, usually metals, are to be removed by cutting. However, before starting work, a little time should be taken to choose a good quality hacksaw frame and to select the most appropriate blade.

Hacksaw frame
The hacksaw frame shown below is adjustable to allow various blade lengths to be fitted. It has a comfortable handle and a rigid construction

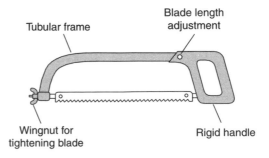

Tubular frame

Blade length adjustment

Wingnut for tightening blade

Rigid handle

Selecting the hacksaw blades

Hacksaw blades are available in a range of **lengths**, **pitches** and **materials**. The correct blade should always be selected. The selection is determined by the size, material and access to the workpiece that is being cut.

Length of the hacksaw blade

The blade length is measured between the two small holes at the ends. When selecting a blade for an adjustable hacksaw frame, choose a short blade if strength is important. However, a short blade would not be suitable for wide workpieces as the hacksaw frame would catch the workpiece. Some hacksaw frames are not adjustable, so you must use a standard length (300 mm) blade.

Pitch of hacksaw blade

The pitch is the teeth spacing. It is specified by the pitch of the teeth. There are three pitches generally available, namely 1.25 mm, 1 mm and 0.5 mm.

Two general rules to follow when selecting the pitch of a hacksaw blade are as follows:

Rule A

Cut with as many teeth as possible without clogging the blade, so:

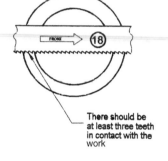

There should be at least three teeth in contact with the work

- use **fine** pitch blades for **hard** metals (to get more teeth doing the work)
- use **coarse** blades for **soft** metals (because there is space for the chips).

Rule B

Always keep three teeth in contact with the work at all times, so:

- use **fine** pitch blades for **thin** sections (to maintain three teeth in contact with the workpiece)
- use **coarse** pitch blades for **thick** metals (because there is space for the chips).

1.25 mm pitch
Soft materials and thick materials

1 mm pitch
General purpose

0.75 mm pitch
Hard materials and thin materials (including thin wall tube)

Hacksaw blade materials

- Flexible – Suitable for cutting mild steel and soft metals. Flexible blades are more durable, so they are used when access to the work is restricted and the blade may bend.
- All hard – Very hard high-speed steel (HSS) but also quite brittle. All hard blades are used when there is good access to the work and it is firmly secured in the vice.
- Bimetal – HSS cutting edges are welded onto a carbon steel backing, making them the most useful blade. Bimetal blades are good for most applications.

Flexible

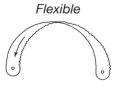

All hard

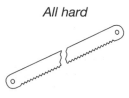

Bimetal

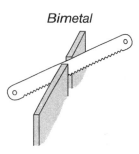

Hacksaw blade tensioning

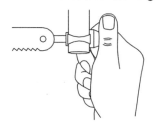

Mounting the blade. The blade is mounted between the two pegs with its teeth facing away from the handle so that the cutting takes place on the forward stroke. The blade is tightened with the winged nut at the front of the frame. The blade should be taut and the frame not over stressed. Usually **three full turns** of the winged nut provides sufficient tension to hold the blade at the correct tension.

Using hacksaws. When using a hacksaw, it is important that the workpiece is first secured in a vice and that the saw handle is held vertically with its frame in line with your arm. If you do not hold it properly, it becomes unsafe and the blades may get broken, material may not be sawn straight and inaccurate sawcuts could be made.

Generally, good and safe techniques are listed as follows:

- Wear safety glasses; if blade breaks, splinters could go anywhere.
- Select the correct blade.
- Check the blade for damage before starting work.
- Mount blade the right way round and apply tension.
- Take care to cut on the correct side of the line.
- Securely arrange the work in the vice so that the sawcut is vertical.
- Cut near to the vice so that the work will not vibrate.
- Stand with your arm and the saw's frame in a straight line.
- Stand so that your feet are apart and you are steady.
- Work steadily with a stroke rate of approximately 40 strokes per minute.
- Use the full length of the blade.
- The blade can be turned through 90° for sawing long pieces.

Usual method of holding a hacksaw

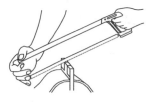

Method of holding a hacksaw for cutting extra long pieces

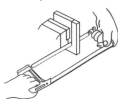

A junior hacksaw

And remember. . .

If the blade does get broken, it is best to finish the cut with a **worn blade**; a new one will jam in the previous slot and will be at a higher risk of breaking.

Junior hacksaw – The junior hacksaw is sometimes handy if a material that must be cut is inaccessible to full-sized hacksaw frames. A junior hacksaw has a frame that is often made from one piece of spring steel. The blades are usually the flexible type and have a fine pitch (32 tpi).

Care of hacksaws
- Store hacksaws with the blade tension slackened.
- Keep hacksaws dry to stop them getting rusty.
- Hang the hacksaw in a rack.
- Protect the blade from being knocked.

Drilling on the bench drilling machine

Bench drilling machine

The bench drilling machine is a light machine used to drill small holes in workpieces when limited accuracy is required. Bench drilling machines are usually mounted on top of workshop benches. A drill chuck with a capacity of approximately 13 mm (1/2″) is usually mounted in the machine's spindle. The drill is secured in the chuck with a chuck key. A small work table measuring approximately 300 mm square is clamped to the machine's pillar and its height is set and locked into position with a clamp.

Bench drilling machine

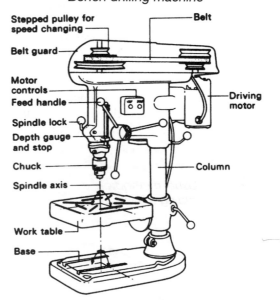

Stepped pulley for speed changing — Belt
Belt guard —
Motor controls —
Feed handle —
Spindle lock —
Depth gauge and stop —
Chuck —
Spindle axis —
Work table —
Base —
Driving motor
Column

The speed of the drill relates to the size of the drill and the metal being cut. If the drill rotates too fast, the drill could 'burn out' owing to excessive friction; if it rotates too slow, time would be wasted. Usually, five speeds are available. The speed is altered by manually changing the position of the belt, so different sized pulleys are engaged.

Refer to Appendix VII. Look up the correct spindle speed for the size of drill/material-type combination.

Setting the spindle speed

Safety – First **isolate** the machine by switching off the electric supply.

Set the spindle speed by removing the belt guard and manually changing the belt's position on the pulleys.

Always move the belt from a larger to a smaller diameter pulley first.

Work holding

Workpiece held in a small machine vice

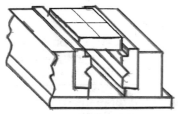

The workpiece must be held securely while it is being drilled. The vice both grips and supports the work. Workpieces are usually mounted on a small vice like the one shown. Mount the workpiece and make sure the drill will 'break through' (i.e. when the drill begins to come out of the other side of the workpiece) without touching the vice. Some small machine vices have a step on their jaws to support flat workpieces. There are also sometimes vees in the jaws to hold round materials vertically or horizontally. Make sure that there is space under the workpiece and that the drill will not hit a vice jaw.

The vice's handle must be firmly held by the fitter to prevent it from rotating during the drilling process; for large holes, the vice must be clamped firmly to the machine table.

Drilling the hole

The workpiece is positioned exactly underneath the drill point.

Safety note! Position the chuck guard around the cutting area and fix it into position before the spindle is turned on. The drill is manually fed through the workpiece with the large three handled handwheel. Always feed slowly for the last bit while the drill is breaking through. The depth of the hole is displayed by viewing a calibrated rod on the front of the machine. Particular care must be taken when the drill is 'breaking through'.

Drilling with the guard in place

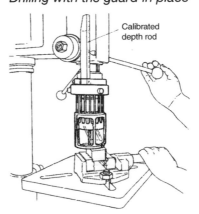

Calibrated depth rod

Never work on any machine until you have been shown how to operate it properly by a suitably qualified person.

Your supervisor will arrange for you to be shown how to operate a bench drill. He will also explain the safety requirements, emergency stop procedure and introduce you to the following features:

- Location of the workshop emergency stop buttons.
- Location of the bench drill's isolator switch.
- Drill chuck and chuck key.
- Spindle stop and start control.
- How to change spindle speed.
- The hand-wheel for feeding the drill.
- Work table and vice.
- Parallel bars for work setting.

If you do not understand what you have been shown or if you are unsure how to proceed you must *ask*!

EXERCISE 8.2   Tee slot cleaner

Mark out and make the tee slot cleaner using the procedure list the tools and equipment selected in the box below. Your completed work should be within the drawing limits, so you must take special care with the two dimensions with limits.

On completion of the task, carefully measure your work and write in the table the actual sizes of your finished workpiece, noting any errors outside the permissible limits.

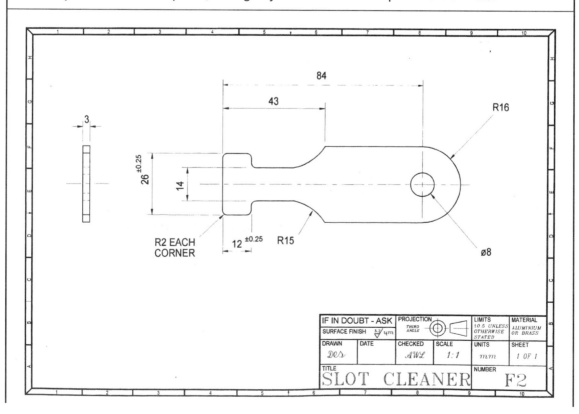

Planning:

Procedure	Tools and equipment selected
1. Acquire material and check for size. 2. Mark out all centre lines and edges. 3. Dot punch the profile and hole/radius centre. 4. Scribe the arcs with dividers. 5. Rough cut the profile with a hacksaw. 6. Finish file the profile. 7. Drill the hole. 8. Stamp name. 9. Deburr and polish. 10. Clean the work area.	

Complete the tables with the correct information:

Saw blade type and pitch used:	How checked before use:
Drill spindle speed selected:	How selected:

List here the safety **procedures** to be observed while operating the bench drill:

1.

2.

3.

Sketch:

Make a sketch of the pulley arrangement on a bench drill showing the spindle speed for each belt position.

Finished inspection report:

Component dimensions (mm)	Limits:	Actual size:	Error:
26 length	26.25 25.75		
12 width	12.25 11.75		
Overall length 100	100.5 99.5		
16 radius	To fit radius gauge		
2 radius	To fit radius gauge		

Results:

Is the workpiece to the drawing specification?

YES	NO

If completed workpiece is below drawing specification, list here the reasons for any errors, stating clearly how the errors will be avoided in future.

Start time and date:	End time and date:	Time taken (hours):

Witness testimony

I confirm that the tee slot cleaner was made within the stated limits and that
. . . . cleared the work area on completion of the task and surplus material was appropriately disposed of.

Signed . . . . Job title . . . . Date . . . .

Hole cutting tools

There are a wide range of cutting tools available for use in drilling machines. The cutting tools described below are used to prepare, rough out and finish holes accurately to size as well as cut special profiles. Always take care when handling hole cutting tools as they have sharp edges and can cause nasty cuts and scratches.

Drills

- **Centre drills** – Double-ended cutting tools are held in a drilling machine's chuck. They are used to produce accurately positioned small centre holes that guide the larger twist drills (described below).

  Centre drills' points and stiff body make them unlikely to wander. Use a high speed to start the hole and drill to a depth of approximately half way down the 60° taper using **gentle** pressure.

Centre drill

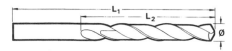

Courtesy of Hydra Tools International

- **Twist drills** – Available in a wide variety of sizes from approximately 0.4 mm (1/64″) to 50 mm (2″). They are used for cutting holes in workpieces. Most twist drills are made from HSS. A drilled hole is quick to produce and can be used for a wide variety of functions, usually for fasteners to go through and for starting threads. Two types of twist drill are made as described below:

  (a) **Parallel shank 'jobber' drills** (up to approximately ø13 mm). These drills have parallel shanks and are made out of HSS. They are held in a drill chuck, and the size of the drill is normally engraved on the shank.

Parallel shank jobber drill

Courtesy of Hydra Tools International

  (b) **Morse taper shank drills** (from approximately ø10 mm to ø50 mm). The Morse taper shanks are used for accurate location and positive drive. The size of the morse taper depends on the size of the drill. There are eight sizes of morse tapers made, numbered from 0 to 7. Size numbers 1, 2 and 3 are most common.

Morse taper shank drill

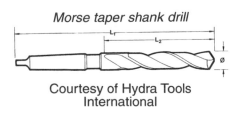

Courtesy of Hydra Tools International

Reamers

Reamers are for accurately finishing predrilled holes to size and shape. They also produce a better quality surface finish than drilled holes. Reamers are supplied as hand or machine type, both types having multiple cutting edges and made from HSS.

- **Hand reamers** – (not used in drilling machines). These are hand tools used for enlarging predrilled holes accurately to size. Hand reamers are gripped in a tap wrench by the square on their shank.

They are smeared with a special cutting fluid, held square to the workpiece and steadily turned and manually pushed into the hole. Hand reamers are tapered for approximately one third of the length of their cutting edges end to assist alignment. This taper makes them unsuitable for finishing blind holes (holes that do not go right through the workpiece).

Hand reamer

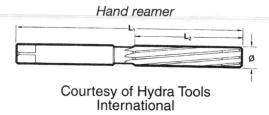

Courtesy of Hydra Tools
International

- **Machine reamers** – Held in a machine tool, this type of reamer has a chamfer on its leading edge because alignment is provided by the machine's construction. Most have a Morse taper shank.

Machine reamer

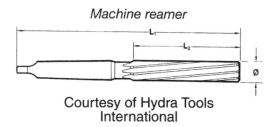

Courtesy of Hydra Tools
International

When setting a machine reamer in a drilling machine, the spindle speed should be set at half the speed for the equivalent size of drill and special cutting fluid applied.

Hole-forming tools

A hole-forming tool is a cutting tool used to change the shape of a predrilled hole. This is usually required to enable the hole to accommodate a fastener (screw or rivet). Most hole-forming tools are made from HSS and have a straight or a Morse-tapered shank. Three of the most common types of hole-forming tools are described below:

- **Countersink** – Usually have a 90° included angle; they are used for flaring out the edge of holes for countersunk head screws or rivets. Also used for making a small countersink when riveting.

- **Counterbore** – Used for making a recess so that cap head screws can fit flush when tightened. To use a counterbore, first drill a pilot hole to fit the counterbore's pilot, then follow by opening out the hole to the required depth with the counterbore.

- **Spot-facing tool** – Used to clean up uneven surfaces around holes.

A spot-facing tool is used when fasteners would otherwise tighten up against a rough surface.

Countersinking tool

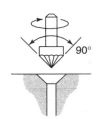

Counterboring tool

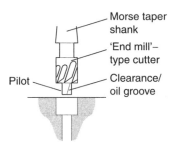

Morse taper shank

'End mill'– type cutter

Pilot

Clearance/ oil groove

Spot-facing tool

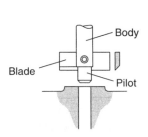

Body

Blade

Pilot

Tool-holding devices used in drilling machines

All drilling machines are equipped with a device for securely holding the cutting tools. In nearly all cases, this is either a drill chuck or a Morse taper in the machine's spindle.

A drill chuck and key

- **Drill chuck** – A drill chuck is a device used to hold parallel shank cutting tools in drilling machines. Most drill chucks have a capacity of approximately 13 mm (1/2″) and are tightened with a special 'chuck key'. They are very quick to operate. When using a drill chuck to hold a parallel shank drill, always make sure that there are no burrs on the sides of the drill's shank or it will not run true.

Safety Note! Always remove the chuck key before starting the machine.

- **Morse-tapers and sleeves** (sockets) – Many machine tools are manufactured with a Morse tapered hole in their spindle. The tapered hole is accurately machined to ensure that the cutting tool rotates properly. Morse tapers can transmit much more power than chucks and are generally used for larger drills and reams.

Cutting tools inserted into a Morse taper will be held securely. If the Morse taper of the cutting tool is not the same as the machine's, a Morse taper sleeve must be used to make up the difference between the two tapers.

Morse taper sleeve and drill drift

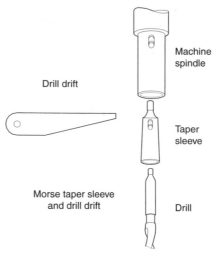

A drill drift is used to extract Morse-tapered tools from tapered spindles and sleeves. It fits into a slot in the spindle and is given a sharp tap with a hammer to eject the tool.

The pillar drilling machine

The pillar drill is a floor-mounted, freestanding drilling machine. Pillar drilling machines are used for drilling and reaming large holes accurately in position. They can also be fitted with special tools to allow them to tap (thread) holes. A machine centre or a centre finder can be fitted into the machine spindle.

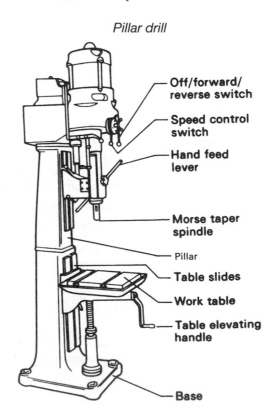

Pillar drill

Features of the pillar drill

- Table equipped with 'tee slots' for clamping the work or work-holding device into position.
- Large range of spindle speeds.
- Means of power feeding the drill through the work.
- Morse taper in their spindle's nose that allows large hole-cutting tools to be fitted directly into the spindle.
- Gear drive train to spindle (not belts).
- Rigid and robust design.
- Powerful electric motor.
- Coolant supply to increase the life of the cutting tool and improve the work's surface finish.

Using a pillar drill

(a) Carefully mark out the centre of the hole. The hole's centre point is aligned with the spindle's centre line using a machine centre or a centre finder and the work in secured in position. Time spent lining up the work initially is well spent as the hole will be more accurately positioned.

(b) Securely clamp the work in position. If it is to be held in a vice, the workpiece should be mounted on parallels, so that when the drill has passed through the work, it does not damage the machine table.

(c) Raise the spindle and insert a drill chuck. Fit a centre drill into the chuck and set the spindle speed accordingly (see Appendix VII).

(d) Turn the spindle on and carefully feed the centre drill down into the work. It should be fed to a depth of approximately 2/3 the depth of the 60° section of the centre drill.

Drilling a large hole. Note the clamps to hold the work securely

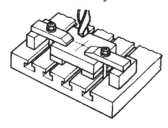

(e) Keep the work clamped in the same position. Remove the centre drill from the chuck and fit a drill. Its position is exactly above the centre drilled hole, so there is no lining up to do.

(f) The spindle speed must be reset, then the hole can be drilled with power feed and at the correct speed. (See Appendix VII for drilling feedrates.)

(g) Substitute the drill for a larger one; if it is outside the range of the chuck, a drill with a Morse taper shank can be fitted directly into the spindle. Again, it is necessary to reset the spindle speed.

(h) Feed the drill through the predrilled hole to open out the hole (increase its size).

(i) Should it be necessary, a machine reamer can be inserted into the spindle and the speed adjusted as appropriate (usually a speed of half the calculated speed for drilling is correct for reaming). The reamer will accurately finish the hole to size and ensure that it is exactly round.

Never **work on any machine until you have been shown how to operate it properly by a suitably qualified person.**

Your supervisor will arrange for you to be shown how to operate a pillar drill. He will also explain the safety requirements, emergency stop procedure and introduce you to the following features:

- Location of the pillar drill's isolator switch
- Drill chuck and chuck key

- Spindle stop and start control
- How to change spindle speeds and feed
- The handwheel for feeding the drill and power feed engagement
- Work table, vice and parallel bars for work setting.

If you do not understand what you have been shown or if you are unsure how to proceed you must *ask*!

EXERCISE 8.3 Depth gauge body

Mark out and make the depth gauge body using the procedure indicated. Complete the tools and equipment selected column. Take special care with the two dimensions with limits. Your finished work should be within the drawing limits.

On completion of the task, carefully measure your work and write in the table the actual sizes, noting any errors outside the permissible limits.

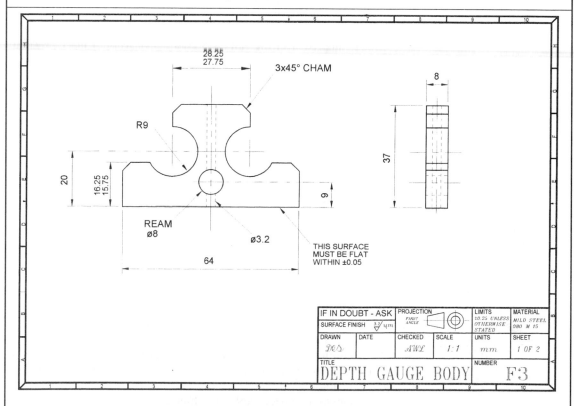

Planning:

Procedure	Tools and equipment selected
1. Acquire material and check for size.	
2. Mark out all lines, centre lines and edges.	
3. Centre drill, drill and chamfer each ø20-mm hole in turn.	
4. Carefully drill the ø3.2-mm hole starting at the top.	
5. Drill and ream the ø8-mm hole.	

Procedure (cont.)	*Tools and equipment selected (cont.)*
6. Rough saw the profile. 7. File base flat. 8. File profile and chamfer as per drawing. 9. Stamp name, deburr and polish. 10. Clean the work area.	

Complete the tables with the correct information:

Protective clothing needed while operating pillar drill:	Reason:
Drill spindle speeds selected	How selected:

Write here about the guarding devices fitted to your pillar drill and how these are checked:

1. Emergency stop:

2. Cutting tool:

3. Belts/pulleys:

Sketch: Make a sketch here to show how your workpiece was securely held while drilling the ø20-mm holes:	

Finished inspection report:

Component dimensions (mm)	Limits:	Actual size:	Error:
16-mm shoulder	16.25 15.75		
28-mm width	27.75 28.25		
Base flat	'Feeler tight' within 0.05 mm and 'Blue' evenly		
Chamfer 3 × 45°	3.5 2.5		

Results:

Is the workpiece to the drawing specification? | YES | | NO |

If completed workpiece is below drawing specification, list here the reasons for any errors, stating clearly how the errors will be avoided in future.

Start time and date:	End time and date:	Time taken (hours):

Witness testimony

I confirm that the drill drift was made within the stated limits and that
.... was seen using appropriate personal protective equipment
while operating the drilling machine.

Signed .... Job title .... Date ....

Cutting internal threads

Internal threads are cut using thread-cutting tools called 'taps'. Taps are precision-cutting tools made from HSS. The tap is screwed into a predrilled hole. As it goes into the hole its sharp cutting edges cut into the sides of the hole, cutting a thread profile.

Tapping can be carried out:

- by hand at a bench using a 'tap wrench' to manually turn the tap (the process is described below)
- machine tapping (in a drilling machine, lathe or milling machine), uses a 'tapping head' (this process is described on page 303).

Hand taps

Hand taps are supplied in sets of three. The names of each tap is shown below together with its uses:

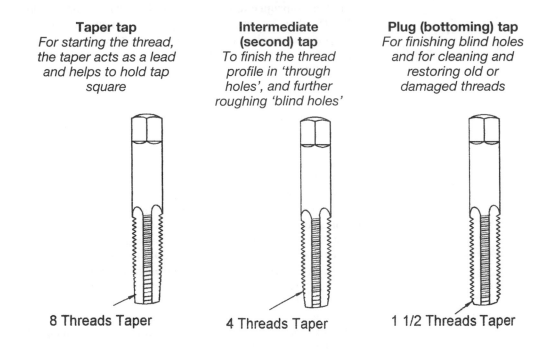

Taper tap
For starting the thread, the taper acts as a lead and helps to hold tap square

Intermediate (second) tap
To finish the thread profile in 'through holes', and further roughing 'blind holes'

Plug (bottoming) tap
For finishing blind holes and for cleaning and restoring old or damaged threads

8 Threads Taper

4 Threads Taper

1 1/2 Threads Taper

Tap wrenches

Tap wrenches are used to hold taps. The Vee shape in the tap wrench's jaws grip the square at the end of the tap's shank. It is important to select a tap wrench that is a good fit on the tap or else the fitter cannot control the tap properly. A range of designs of tap wrench are available. When selecting a tap wrench, the length of the wrench should be minimal to lessen the chances of tap breakage and to increase 'feel'.

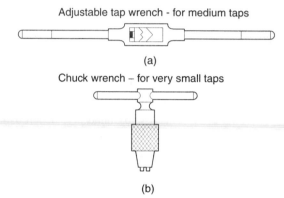

Two types of tap wrench – adjustable and T-handle types

Adjustable tap wrench - for medium taps

(a)

Chuck wrench – for very small taps

(b)

Using taps to cut internal threads

The process of hand-tapping is as follows:

(a) Find the **appropriate** size of hole to drill by referring to Appendix III. The thread's details may be given on an engineering drawing or you may need to measure the screw. If you are measuring a thread, use screw pitch gauges to assess the pitch and a micrometer to find its diameter.

(b) Carefully mark out, centre punch and drill the hole the 'tapping size'.

(c) Select suitable size tap wrench to turn the tap (see tap wrenches).

(d) Smear some 'tapping compound' onto each tap's surface before use. This compound helps the tap to cut by lubricating the cutting edges.

(e) Hold the work firmly in a vice.

(f) Mount the taper tap in the wrench.

(g) Place the tapered end of the tap into the hole.

(h) Hold the tap vertically and square to the hole's axis.

(i) Rotate the tap and press down at the same time.

(j) Once the tap starts to cut, turn it half a turn forwards, then relieve it a quarter turn backwards, advance again half a turn and relieve quarter turn and so on until the thread depth is achieved. (The relieving breaks up the chips so that the tap does not become jammed.)

(k) Remove the tap by unscrewing the tap wrench and change the tap for the intermediate tap.

(l) Apply a little cutting compound and screw this tap into the hole to finish the thread form.

- Blind holes must be retapped with the plug tap to make sure that the thread's full form extends to the bottom of the hole.

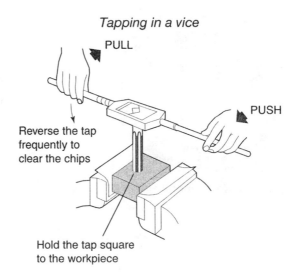

Tapping in a vice

PULL

PUSH

Reverse the tap
frequently to
clear the chips

Hold the tap square
to the workpiece

Care of taps

Always take care with taps. They are precision made and very brittle.

- If subjected to excessive force, taps will break. Broken pieces of taps are very difficult to remove from inside holes. A broken tap will probably scrap your workpiece.
- Always break up swarf by relieving the tap quarter of a turn, a turn backwards for each half turn forwards.
- Always keep taps as a set – taper tap, intermediate tap and plug tap.
- Taps should be cleaned after use and stored in their box.

Cutting external screw threads

External threads are cut using tools called 'dies'. Dies are precision cutting tools made from HSS. The die is screwed onto the end of a chamfered workpiece; as it goes onto the workpiece, its sharp cutting edges cut into the sides of the bar forming a thread profile.

External screw thread cutting can be carried out:

- by hand, at a bench using a 'die stock' to manually turn the die; the process is described below
- by hand on a centre lathe using a 'die stock' to manually turn the die, the process is described in the Turning chapter of this book.
- on a centre lathe using a special 'die head'. This process is also described later in this workbook.

Types of die used for hand thread cutting

The two main types of die are shown below together with their uses:

Circular split die
The most common type of die. It is used for cutting external threads on bars

Courtesy of Kennametal Hertel EDG Ltd

Hexagon die nut
This type of die is screwed onto damaged or old threads with a spanner to clean and restore its profile

Courtesy of Kennametal Hertel EDG Ltd

Die stocks

Die stocks are used to hold split button dies. They have a bored recess that holds the die.

Four common sizes of bore:

$$\frac{13}{16}'' \qquad 1'' \qquad 1\frac{5}{16}'' \qquad 1\frac{1}{2}''$$

The size of die stock selected depends on the diameter of the die. Die stocks have three small screws on their edge for adjusting the size of the die (see below). When selecting a die stock, its length should be minimal to lessen the chances of chipping the die and increase the 'feel'.

A die stock

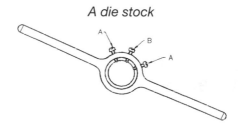

Using dies to cut external threads

The process is as follows:

Chamfered workpiece

90° lead

(a) Select the correct size material; its diameter must be the outside diameter of the finished thread.
(b) Chamfer the end of the workpiece with a file and mount it vertically in a vice.
(c) Select a suitable size die stock to mount the die (see die stocks above).
(d) Insert the die into the stock. The two screws labelled 'A' are slackened and screw 'B' is tightened to open the split die to its maximum size.

Adjusting the die by tightening screw 'B'

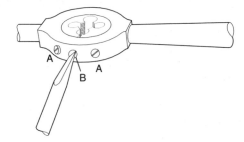

(e) Smear some 'tapping compound' (cutting paste) onto the die's surface before use. This compound helps the die to cut by lubricating its cutting edges.

(f) Cut the thread:

Cutting a thread with a stock and die on a vice

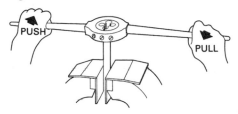

- Hold the die stock square to the workpiece.
- Push firmly on the end of the die stock and turn it clockwise to screw it onto the workpiece.
- Check for squareness after the first two full turns.
- Turn the die stock quarter turn back after each half turn forwards to break up the swarf.
- Continue until full length is reached.
- Remove the die by unscrewing it from the workpiece.

Check the thread:

(a) Examine the surface.
(b) Compare it with screw pitch gauges.
(c) Try on a nut or thread gauge.
(d) If the nut is tight or will not go on:

- slacken screw 'B' slightly
- tighten screws 'A'
- apply a little more cutting compound
- screw the diestock onto the work again to finish its profile.

Care of dies

Always take care with taps, they are precision made and very brittle.

- If subjected to excessive force, dies will break.
- Always break up swarf by relieving the die quarter of a turn, a turn backwards for each half turn forwards.
- Dies should be cleaned after use and stored in their case.

EXERCISE 8.4   Alignment tool handle

Complete a procedure plan to make the alignment tool handle. Complete the tools and equipment column. Take special care with the thread to ensure that it is square to the axis of the body. Get your plan checked by your supervisor and then make the alignment tool handle to your plan.

On completion of the task, carefully measure your work and write in the table the actual sizes of your finished workpiece, noting any errors outside the permissible limits.

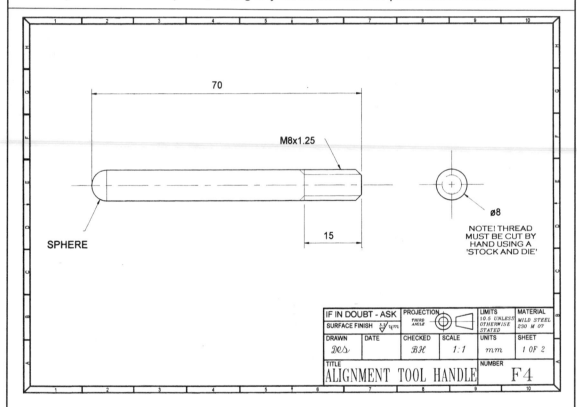

Planning:

Planned procedure	Tools and equipment selected
1.	
2.	
3.	
4.	
5.	
6.	
7.	
8.	
9.	
10.	

Complete the tables with the correct information:

Size diestock used	How measured:
Possible defects in external threads cut with a diestock:	How checked for this defect:
Method of checking thread:	How selected:

List here the safety **procedures** to be observed to avoid dermatitis:

1.

2.

Sketch:

Sketch a metric external screw thread profile showing clearly the thread's

- pitch
- flank angle
- major diameter
- minor diameter.

Finished inspection report:

Component dimensions (mm)	Limits:	Actual size:	Error:
Overall length 70	70.5 69.5		
Thread length 15	15.5 14.5		
Sphere R4	*To fit radius gauge*		

Thread profile	*To fit 1.25 thread pitch gauge*		

Results:

Is the workpiece to the drawing specification?	YES	NO

If completed workpiece is below drawing specification, list here the reasons for any errors, stating clearly how the errors will be avoided in future.

Start time and date:	End time and date:	Time taken (hours):

Witness testimony

I confirm that the alignment tool handle was made within the stated limits and that. . . . . .cleared the work area on completion of the task, disposing of waste and surplus materials safely.

Signed . . . . .Job title . . . . Date . . . .

Shaping machine

A shaping machine is a machine tool that is sometimes available to fitters in engineering workshops. Shapers, as they are commonly known, are simple machines to operate and quick to set up. They can be used to produce flat surfaces in either horizontal, vertical or angular planes on workpieces. The Ram carries a cutting tool to and fro over the work and the cutting tool removes material from the work's surface as it traverses forward over the work, hinging away from the work on the return stroke. For each cutting stroke, the work can be automatically fed across the tool's path so that the next stroke will remove more material. The stroke length and the feedrate can be adjusted to suit the workpiece's dimensions and the required surface finish.

The shaping machine

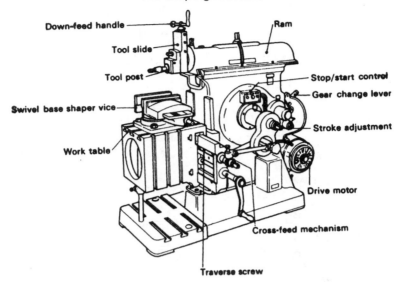

Never **work on any machine until you have been shown how to operate it properly by a suitably qualified person.**

Your supervisor will arrange for you to be shown how to operate the shaping machine if you have one in your workshop. He will also explain the safety requirements, emergency stop procedures and introduce you to the following features:

- Location of the isolator switch
- How to set the stroke length and position
- Stop and start control
- How to set speeds and feed
- The handwheel for moving the cutting tool
- The method of feed engagement
- Work table, vice and parallel bars for work setting.

If you do not understand what you have been shown or if you are unsure how to proceed, you must *ask*!

EXERCISE 8.5   Tool maker's clamp jaws

Make the tool maker's clamp jaws using the procedure below and list the tools and equipment selected in the box below. Your finished work should be within the drawing limits. (This workpiece was marked out in Exercise 7.5.)

On completion of the task, carefully measure your work and write in the table the actual sizes, noting any errors.

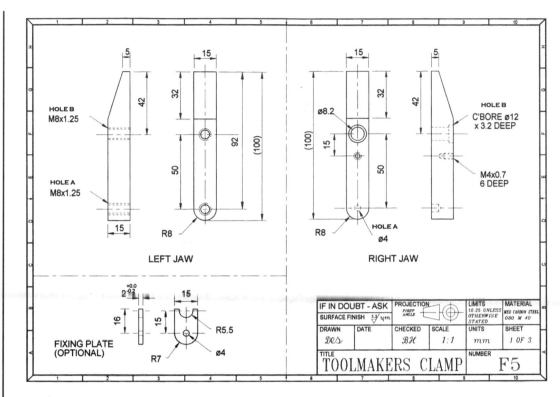

LEFT JAW

RIGHT JAW

HOLE B
M8x1.25

HOLE A
M8x1.25

HOLE B
C'BORE ø12
x 3.2 DEEP

M4x0.7
6 DEEP

HOLE A
ø4

R8

FIXING PLATE
(OPTIONAL)

R5.5

R7

ø4

IF IN DOUBT - ASK	PROJECTION FIRST ANGLE		LIMITS 10.25 UNLESS OTHERWISE STATED	MATERIAL MED CARBON STEEL 080 M 40
SURFACE FINISH 3.3/ μm				
DRAWN DCA	DATE	CHECKED BH	SCALE 1:1	UNITS mm	SHEET 1 OF 3
TITLE TOOLMAKERS CLAMP				NUMBER F5

Planning:

Planned procedure	*Tools and equipment selected*
1. Acquire material and check for size. 2. Mark out all lines, centre lines and edges on left jaw. 3. Hold the two jaws together with a pair of tool maker's clamps. 4. Drill holes A through M8 × 1.25 hole at the tapping size. 5. Drill holes B through the left jaw and 5 mm deep into the right jaw with a ø5-mm drill. 6. Separate the parts. 7. Redrill, counterbore and tap each jaw as appropriate. 8. Cut angular faces with shaper, file radius at ends. 9. Stamp name, deburr and polish. 10. Clean the work area.	

Record of discussion with your supervisor about the planned procedure.

Signed by Supervisor . . . .   Date . . . .

Complete the tables with the correct information:

Selection of tools and equipment:	

1. **Both** tapping size drills:
  M4 × 0.7:
  M8 × 1.25:	

2. Length of tap wrenches used:
  For M4 × 0.7 thread:
  For M8 × 1.25 thread:	

3. Name of tapping compound used:	

Sketch:

Sketch your set-up for holding the two pieces together while drilling the two pairs of holes exactly in line.

Finished inspection report:

Component dimensions (mm)	*Limits:*	*Actual size:*	*Error:*
Hole centres 50	*50.25* *49.75*		
C'bore depth 3	*3.45* *2.95*		
Check tapped thread form for excessive tearing:	*Visual check*		

Use appropriate gauges to check pitch of thread:	1.25-mm and 0.7-mm thread pitch gauge		

Results:

Is the workpiece to the drawing specification?	YES	NO

If completed workpiece is below drawing specification, list here the reasons for any errors, stating clearly how the errors will be avoided in future.

Start time and date:	End time and date:	Time taken (hours):

Witness testimony

I confirm that the tool maker's clamp jaws were made within the stated limits and that. . . . . used appropriate ppe, correctly used machine guards and safety mechanisms and cleared the work area on completion of the task.

Signed . . . . Job title . . . . Date . . . .

Reaming

A reamer is a cutting tool used to **accurately** finish predrilled holes to size. Reamers have multiple cutting edges and are made from high-speed steel. They are designed to remove only **small** amounts of material and leave a **smooth surface finish** of approximately $0.8\,\mu m$, much better than an ordinary drill.

There are two types of reamers:

- Hand reamers.
- Machine reamers.

Reamers were introduced in the section on hole-forming tools (p. 222).

The two processes for reaming holes are as follows:

Hand reaming

(a) Accurately mark out and predrill the hole's position. The size of the hole is important – refer to Appendix X to select the correct drill size.

Note. The size of a drilled hole is more likely to be accurate if it is first pilot drilled and then redrilled to the required size.

(b) Hold the work firmly in the vice with the hole's axis vertical.

(c) Apply a smear of cutting fluid to the reamer's body.

(d) Hold the square on the reamer's shank in a tap wrench. The wrench should hold the reamer firmly.

Hand reaming in a bench vice

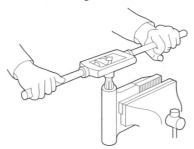

(e) Hold the tap wrench in the palm of your hand and locate the end of the reamer in the hole. It will go into the hole part way as the first quarter of a hand reamer's length is tapered.

(f) Hold the reamer square to the hole's axis, rotate it clockwise and press it into the work at the same time.

(g) Remove the reamer by continuing to turn it clockwise while withdrawing it from the hole.

Machine reaming

Generally, machine reamers are more accurate than hand reamers.

(a) Mark out the workpiece and clamp it accurately in position on a drilling machine's table.

(b) Centre drill and predrill the workpiece to the correct drill size after referring to Appendix X.

(c) Remove the drill and insert the machine reamer in the machine's spindle.

(d) Apply a little 'tapping compound' to the reamer if coolant is not available on the drilling machine.

- If a Morse taper shank reamer is used, it might be necessary to use a Morse taper sleeve. Remember that the machine holds the reamer straight and square to the workpiece.

- General practice is to use half the drilling speed and double the recommended feedrate for the equivalent drilled hole – refer to Appendix X (reaming allowances).

- The machine holds the reamer square to the hole. Turn on the spindle and feed the reamer slowly through the hole.

- Machine reamers cut on the chamfered leading edge of their teeth. The sides of the reamer guide the body straight down the hole and prevent it from running out of true.

- Apply a generous supply of coolant, if it is available on the machine; If coolant is not available apply some tapping compound to the reamer's cutting edges.

Work held in a machine vice ready for reaming

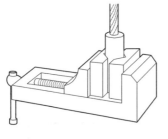

(e) Withdraw the reamer, with the drilling machine still running, immediately after 'breaking through' or reaching the required depth.

EXERCISE 8.6　Drill stand plates

Write a plan to make the drill stand plates using the planning list. List all tools and equipment to be used in the task. Get your plan checked by your supervisor and then make the job in the order you planned. Your finished work should be within the drawing limits.

On completion of the task, carefully measure your work and record the actual sizes, noting any errors.

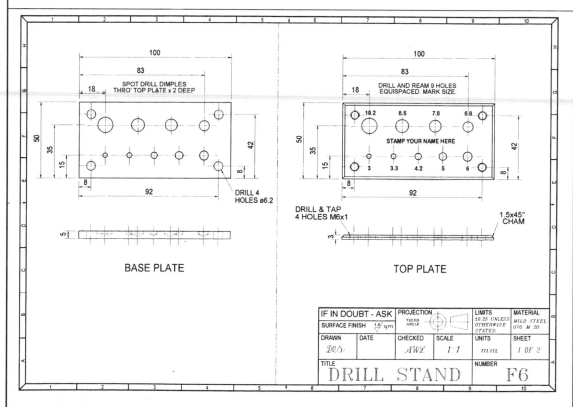

Planning:

Procedure	Tools and equipment selected
1. Acquire material and check for size. 2. 3. 4. 5. 6. 7. 8. 9. 10.	

Complete the tables with the correct information:

The following holes are reamed. Write in the boxes the appropriate drill sizes required and the appropriate spindle speeds for reaming.

Ø3 mm	Ø5 mm	Ø6 mm
Size: Speed:	Size: Speed:	Size: Speed

Type of reamer used (hand or machine):	How checked for suitability to the work:
Name of cutting tool used to machine counter sink:	How checked for suitability to the work:
Cutting fluids used: 1. for tapping: 2. for reaming:	How applied safely to the workpiece:

List here how the machine guards and other safety mechanisms are used while drilling:

1.

2.

3.

Sketch:

Make a clear sketch to illustrate how the depth is set on your drilling machine.

Machine's depth stop

Finished inspection report:

Component dimensions (mm)	Limits:	Actual size:	Error:
Overall width			
Overall length:			
Chamfer:			
Corner hole positions:			
Surface texture:			

Results:

Is the workpiece to the drawing specification?

YES	NO

If completed workpiece is below drawing specification, list here the reasons for any errors, stating clearly how the errors will be avoided in future.

Start time and date:	End time and date:	Time taken (hours):

Witness testimony

I confirm that the drill stand plates were made within the stated limits and that. . . . . .cleared the work area on completion of the task.

Signed . . . . .Job title . . . . Date . . . .

Screwdrivers

Screwdrivers are designed to turn the heads of screwed fasteners. There are various designs and sizes of screwdriver blades and handles.

Selection of screwdrivers

- Blades – The most important feature of a screwdriver is the point of its blade. When selecting a screwdriver, first look at the head of the screw, it should be one of these:
  Types of heads of screws:

Straight
Standard screwdriver tip. Make sure that the tip fits the slot in both length and thickness

Cross/Phillips
Fastener used in a wide variety of applications

Pozi/Supadriv†
Considered to be stronger than the cross head screw. Notice the star

Torx
Used in the motor industry and for electronic components

†Pozi/Supadriv are registered trademarks of the European Industrial Services Ltd.

When selecting a screwdriver, make sure that the size and shape of its point match the shape of the screw's head and are a good fit.

Flat screw- driver tip

Cross/Phillips screw- driver tip

Pozi/Supadriv tip

Torx screw- driver tip

- Handles – The shape and size of the handle affects the amount of 'turning effect' that can be applied to the screw head. A bigger handle is easier to turn than a thin one, and a handle with serrated grips is also easier to turn. A screwdriver handle should be comfortable; different people prefer different styles for their own personal reasons. For electrical work, an insulated plastic handle **must** always be used. Some screwdrivers also have insulation along the length of the blade as well for further protection.

Never:

- hit a screwdriver with a hammer – it spoils the tip and can split the handle
- use a screwdriver to turn screws in a hand-held workpiece because you could stab the palm of your hand with the screwdriver's blade.

Hexagon socket wrenches (Allen keys)

Hexagon socket wrenches are used for turning hexagon socket head screws (known as cap screws) and other special screws. The size of the Allen key is its measurement across the flats of the hexagon (A/F). Made of high-quality tool steel, Allen keys look like bent pieces of hexagonal bar, a screwdriver type is also available. The Allen key is inserted into the socket head screw and turned, thus tightening or undoing the fastener. It is important to use the correct size of Allen key that should fit firmly in the screw's head with no wavering.

Hexagon socket wrench (Allen key)

Types of socket head screw

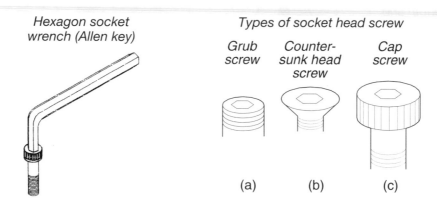

Grub screw	Counter-sunk head screw	Cap screw
(a)	(b)	(c)

Spanners

Spanners are used for tightening and loosening hexagonal and square head nuts and bolts. All good spanners are made of high-quality alloy steel containing chromium and vanadium to increase their strength and toughness. Most spanner manufacturers supply spanners in a range of metric (mm), English (BSW and BSF) and American (A/F) sizes. Spanners are available in a range of styles to suit almost every job.

Open ended
Angled head of spanner allows access to nuts near obstructions

Ring
Grips nut all way round. High strength, cannot be used on pipes. Ends are of different sizes

Combination
One end is a ring; the other is open ended. Both ends fit same-size nut. Provide versatility

Adjustable
Grips and turns almost any size of nut. Can be clumsy. Use when correct size spanner is not available

Courtesy of Britool Ltd

C-spanner

C-spanners are special spanners used for loosening and tightening round slotted nuts. Round slotted nuts are thin in section and evenly balanced. This type of nut is frequently used to hold grinding wheels into position. The C-spanner is hooked into a notch on the edge of the nut and it is turned by the long handle.

C-spanner

Nut requiring a C-spanner

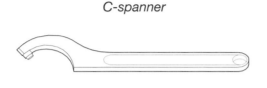

Engineer's socket sets

Socket sets are made up with a variety of wrenches, extension pieces and sockets to fit a range of size nuts. They are strong and versatile. A large range of accessories enable fasteners to be tightened in positions when the end of the nut is accessible. Sockets cannot be used on nuts where access to the end of the nut is restricted because it is close to something or it is on a pipe.

A good quality engineer's socket set

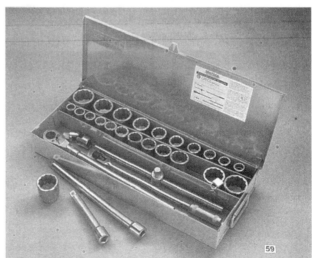

Courtesy of Britool Ltd

Torque wrenches

If nuts are tightened too tight, the shank of a nut or bolt can be permanently stretched and weakened. For many applications it is important to tighten the bolt as tight as possible without causing any weakening. The maximum tightness allowed depends on a number of factors, principally the material and size of the stud/shank. The tightness of a bolt is measured in units of torque called Newton/metres. A torque wrench measures the amount of torque being applied to a bolt. Most torque wrenches use sockets, and are set by turning a screw on the handle to set the required torque. When the correct torque is applied, the torque wrench 'clicks'.

Using a torque wrench

Most bolts tightened to specified torque settings are made out of a special 'high-tensile steel'. There are a number of quality grades of bolts. These grades indicate the strength of the bolt; 8.8 is a fairly popular quality grade of metric bolt. If the manufacturer's data is not available, you can find the correct torque for a metric coarse bolt by referring to Appendix XI, which lists some common torque wrench settings.

To set the torque wrench, turn the adjusting screw on the end of the handle so that the correct torque is shown on the gauge. The socket is put onto the wrench and attached to the bolt. Turn the wrench by the handle and when the correct torque is reached there is a 'click' and the wrench 'gives'. The bolt is now tightened to the correct torque.

Selection of spanners and wrenches:

- Always take care to select a well-fitting spanner or wrench.
- Check for damage or wear before use.
- Select a tool that is most comfortable to use.
- Pull tools towards you (instead of pushing) in case you slip.
- Wipe away any oil from tools before storing.

And never:

- Extend spanners or wrenches with a pipe to increase leverage.
- Tilt open-ended spanners at an angle to the screw head; it is likely to slip.
- Hammer on a spanner or wrench.
- File or grind a spanner to make it fit.

Setting a torque wrench

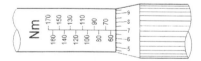

Using a torque wrench

Courtesy of Britool Ltd

Pliers

Pliers are hand tools for gripping irregular-shaped workpieces. Pliers are made of high carbon steel and often have insulated handles. They are available in a range of designs and sizes, the most popular of which are shown and described below:

Selection of pliers

When selecting pliers, it is necessary to be aware of the styles available and their applications. Consider the drawings and uses listed below. The size of pliers refers to the overall length including the jaws and handles; those with longer handles provide more leverage and a tighter grip of the workpiece.

Flat nose
Used only for gripping thin objects. The short flat jaws are serrated inside and hardened

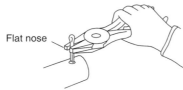

Flat nose

Combination
Jaws incorporate serrated flat grips, splined pipe grips and two types of wire cutters

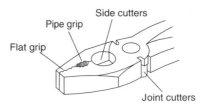

Side cutters

Pipe grip

Flat grip

Joint cutters

Snipe nose
Snipe nose jaws help with handling small objects and give a better view of work

Electricans
Insulation on handles withstands high voltages. Used as combination pliers

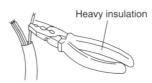

Heavy insulation

Using pliers

- Hold the workpiece in the jaws as close to the hinge as possible to get the tightest grip.
- Hold pliers at the end of the handles to get maximum grip.
- When working with high-voltage electrical appliances, always ensure that the insulation is intact on the handles.
- Do not use pliers for gripping nuts instead of spanners – the pliers could spoil the nut's hexagon.
- When cutting with the pincer blades, always make sure your hand will not get trapped between the handles if the workpiece suddenly breaks.

Pipe wrench (Stillson)

Pipe wrenches are for turning pipes. The serrated jaws of the pipe wrench can be adjusted to fit over a pipe. As you pull on the pipe wrench's handle the jaws grip the pipe, turning it. Pipe wrenches damage the surface of workpiece, but this is not generally a problem with pipework systems.

Safety note! You must not extend the handle of a pipe wrench to increase leverage.

A pipe wrench (Stillson)

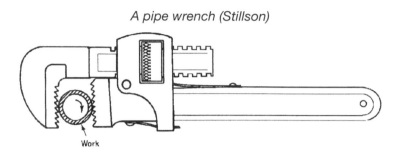

Work

EXERCISE 8.7   Gate valve

Prepare a clean and tidy work area and plan a dismantling sequence for the gate valve (your supervisor may give you another suitable assembly to dismantle for this exercise).

Your dismantled work should be carefully examined and cleaned.
Report any defects that would cause the components to fail.
Ask your supervisor to verify your report and confirm that the workpiece was dismantled correctly.
Reassemble the gate valve and check to make sure that it opens and closes correctly.
The main screws on the body should be tightened to 75 N/m with a torque wrench.

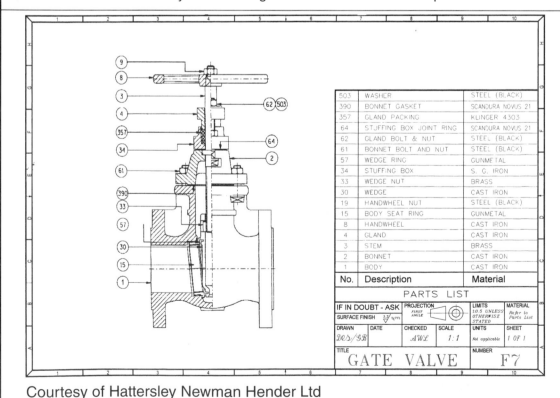

No.	Description	Material
503	WASHER	STEEL (BLACK)
390	BONNET GASKET	SCANDURA NOVUS 21
357	GLAND PACKING	KLINGER 4303
64	STUFFING BOX JOINT RING	SCANDURA NOVUS 21
62	GLAND BOLT & NUT	STEEL (BLACK)
61	BONNET BOLT AND NUT	STEEL (BLACK)
57	WEDGE RING	GUNMETAL
34	STUFFING BOX	S. G. IRON
33	WEDGE NUT	BRASS
30	WEDGE	CAST IRON
19	HANDWHEEL NUT	STEEL (BLACK)
15	BODY SEAT RING	GUNMETAL
8	HANDWHEEL	CAST IRON
4	GLAND	CAST IRON
3	STEM	BRASS
2	BONNET	CAST IRON
1	BODY	CAST IRON

PARTS LIST

IF IN DOUBT - ASK	PROJECTION FIRST ANGLE		LIMITS 10.5 UNLESS OTHERWISE STATED	MATERIAL Refer to Parts List
SURFACE FINISH 3.2 µm				
DRAWN DRS/GB	DATE	CHECKED AWL	SCALE 1:1	UNITS Not applicable	SHEET 1 OF 1
TITLE GATE VALVE				NUMBER F7

Courtesy of Hattersley Newman Hender Ltd

Planning 1 – dismantling:

Planned procedure	Tools and equipment selected
1. Acquire gate valve from store. 2. 3. 4. 5. 6. 7. 8. 9. 10. Clean the work area.	

Refer to Appendix XI regarding torque wrench settings, and when reassembling the gate valve, tighten the major load-bearing bolts to the correct torque.

Complete the tables with the correct information:

	Size	Material type	Torque setting
Load-bearing bolt No. 1			
Load-bearing bolt No. 2			

Planning 2 – assembly:

Planned re-assembly procedure	Tools and equipment selected
1. Acquire gate valve components from store. 2. 3. 4. 5. 6. 7. 8. 9. 10. Clean the work area.	

Torque settings used	
1. Types of pliers used:	
2. Types and sizes of spanners used:	
3. Types of screwdrivers used:	
4. Types of wrenches used:	

Notes on discussions with supervisor about problems or difficulties encountered during the task:

Signed authentic by
Supervisor.............................Date.............................

Write here:

1. The reasons why the work area must be kept and maintained clean and tidy.

2. How any surplus materials were disposed of according to company safety policies.

List below the safety checks made on all hand tools before use:

1.

2.

3.

List the types of materials you have used below under the following categories:

Ferrous	Non-ferrous	Non-metals

Does the gate valve work as it should?	YES	NO

If completed workpiece does not work as it should, write here the reasons for any defects, stating clearly how they could be rectified.

Start time and date:	End time and date:	Time taken (hours):

Witness testimony

I confirm that the gate valve was dismantled, inspected and reassembled correctly with all fasteners and washers replaced and that............................cleared the work area on completion of the task.

Signed . . . . Job title . . . . Date . . . .

Hammers

Hammers are important tools and must be used correctly and safely. Their main uses are:

- tapping marking tools
- riveting
- chiselling
- driving things into position.

Hammers are identified by their shape and weight. Some types of hammer are shown here:

Ball pein	**Cross pein**	**Straight pein**	**Soft faced**
Face of hammer for general purpose. Ball peen for spreading work when riveting	*Pein for riveting in awkward corners. Front face used for general work. Supplied in lighter weights*	*Pein for riveting in awtward corners. Front face used for general work. Supplied in lighter weights.*	*Plugs made of raw hide, aluminium or copper. Used when the component being hit must not be damaged*

Most hammers are available in a range of weights, typically between 225 gm ($\frac{1}{2}$ lb) and 900 gm (2 lb). Bigger hammers than this are called sledge hammers and are used for very heavy work. Hammer handles are made from good quality ash or hickory wood.

Always hold a hammer, firmly at the end, as shown.

Before using any hammer always carry out the following safety checks:

- Head is tight on the shaft.
- Shaft is not split.
- Head is not cracked or mushroomed.
- The handle is dry and not slippery.

Correct method of holding a hammer

Riveting

Riveting is the process of beating out the end of a special pin (called a rivet) or screw to produce a fastening that will not work loose if it is subjected to vibrations.

To form a flush (smooth) rivet, there must be some space for the excess material to fill when it is deformed. Usually a countersink is made in the workpiece for the deformed part of the rivet to fill. The rivet is then inserted into the hole and beaten with the ball peen of a hammer until its end is deformed and fills the countersink. To finish the surface smooth, it must be carefully filed flat with a smooth file.

Round head rivet   Pan head rivet   Countersunk head rivet

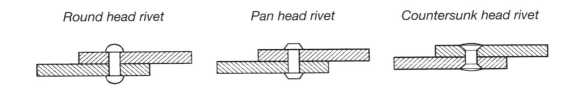

EXERCISE 8.8   Drill stand assembly

On completion of Exercises 8.6 (Fitting F6 (drill stand plates)) and 9.1 (Turning T3 (drill stand legs)), plan a detailed assembly procedure. Assemble the components by riveting the screw threads into the countersunk holes in the top plate. Remember, the legs must be screwed in tightly.

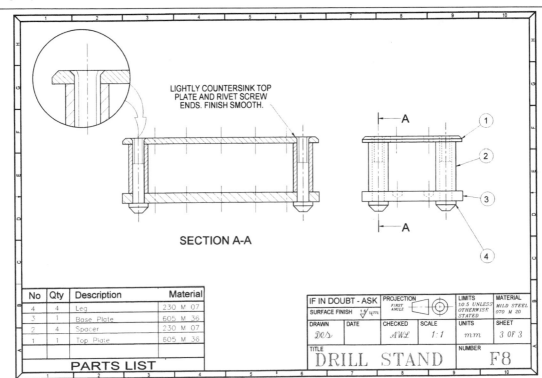

LIGHTLY COUNTERSINK TOP
PLATE AND RIVET SCREW
ENDS. FINISH SMOOTH.

SECTION A-A

No	Qty	Description	Material
4	4	Leg	230 M 07
3	1	Base Plate	605 M 36
2	4	Spacer	230 M 07
1	1	Top Plate	605 M 36

PARTS LIST

| IF IN DOUBT - ASK | PROJECTION FIRST ANGLE | | LIMITS 10.5 UNLESS OTHERWISE STATED | MATERIAL MILD STEEL 070 M 20 |
| SURFACE FINISH 1.6 μm | | | | |

| DRAWN | DATE | CHECKED | SCALE | UNITS | SHEET |
| DES | | AWL | 1:1 | mm | 3 OF 3 |

TITLE **DRILL STAND**

NUMBER **F8**

Planning:

Procedure	*Tools and equipment required*
1. Acquire components and finally check for size. 2. 3. 4. 5. 6. 7. 8. 9. Deburr and polish. 10. Clean the work area.	

Complete the tables with the correct information:

Hammer used for riveting (shape, material and weight):	How checked for safety before starting work:

List safety checks carried out on hammer(s) before starting to rivet the screw ends. 1. 2. 3.	

Sketch:

Sketch a detailed drawing of the tools used to assess the surface texture that is specified as 1.6 μm all over.

Surface texture equipment

Results:

Is the workpiece to the drawing specification?

YES	NO

If completed workpiece is below drawing specification, list here the reasons for any errors, stating clearly how the errors will be avoided in future.

Start time and date:	End time and date:	Time taken (hours):

Witness testimony

I confirm that the drill stand assembly was made within the stated limits and that. . . . . cleared the work area on completion of the task.

Signed . . . . Job title . . . . Date . . . .

Cold chisels

Cold chisels are commonly called chisels. Chisels are used with hammers for removing large amounts of material by hand. A variety of shapes are available, each having its own particular application.

Types of chisel:

Flat
Used for splitting nuts, cutting sheet metal in a vice and general material removal

Cross cut
Used for cutting fine slots, e.g. keyways

Diamond nose
Access to and sharpening of internal corners

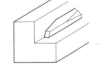

Round nose
Used for cutting rounded grooves for oilways in machine tool slideways

Using chisels

(a) The chisel will need to be inclined at approximately 25°–30° depending on its point angle (60° for general use) and the material you are cutting (see illustrations).

(b) Strike the chisel firmly and repeatedly with a hammer.

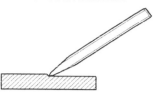

For soft materials

For general purpose

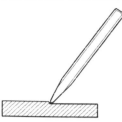

For harder materials

Always:

- Wear safety glasses and a glove when chiselling.
- Use a screen to protect others from your flying chips.
- Hold the hammer properly.

Never:

- Use a chisel with a mushroomed head. (Pieces of 'mushroom' can fly off causing injury.)

Grinding chisel points

The point angle of chisels depends on the material to be cut. Generally, chisels are sharpened to an angle of 60°. For softer materials, slightly more acute (sharper) point angles are appropriate as the chisel would cut more quickly. For harder materials, slightly more obtuse (blunter) point angles would be used as it makes the edge last longer.

Correct chisel point angles for common materials:

Aluminium	Brass	Bronze	Cast iron	Mild steel	High carbon steel	Hard alloy steel
30°	50°	60°	60°	55°	65°	65°

When sharpening a pointed instrument (e.g. a punch, scriber or a chisel), hold it vertically against the grinding wheel at the appropriate angle. Press the instrument lightly against the revolving grinding wheel; the grinding marks will lie away from the edge. This will help to reduce friction in the cutting process and lengthen the edge life of the tool. Take care not to overheat the point while sharpening it, and keep a supply of cooling liquid nearby, for example water or coolant from a machine tool. You can tell if the tool point is too hot as it will turn blue. If this happens, it will need rehardening and tempering.

Sharpening a chisel

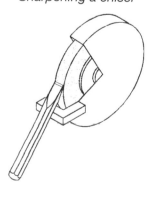

Remember:

While sharpening chisels, turn it round and remove any mushrooming from the chisel's head. Mushrooming is dangerous as pieces can fly off and hit you or others.

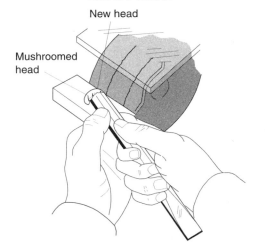

Removing 'mushrooming' from a chisel's shank

New head

Mushroomed head

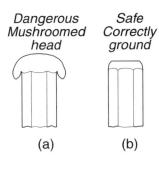

Dangerous Mushroomed head

Safe Correctly ground

(a)   (b)

Ask your Supervisor to show you how to use a grinding machine to grind a chisel's point.

Notice the positions of the machine guards and rests.

Always:

- Wear safety glasses.
- Wear appropriate protective clothing.
- Use the machine screens provided.
- Make sure the gap between the wheel and rest is minimal.

Never:

- Work on a grinding machine of any description without being shown how to operate it.
- Touch a grinding wheel when it's going.
- Try to stop a grinding wheel with anything.

EXERCISE 8.9 Sharpening and using a chisel

Using a safe procedure, sharpen the following tools:

- dot or centre punch
- scriber
- chisel (flat; round nose; diamond point or cross cut).

Use a protractor or point angle gauge to assess the angle of the tools after grinding.

Your finished tools should have the correct lay patterns (grinding marks) and correct angles. On completion of the task, carefully measure your tool's angles and complete the tables below.

60°	90°	55°	20 - 25°
COLD CHISEL	CENTRE PUNCH	DOT PUNCH	SCRIBER

IF IN DOUBT - ASK	PROJECTION		LIMITS	MATERIAL
	N/A		*N/A*	*N/A*
SURFACE FINISH ✓ *µm*				
DRAWN *Des*	DATE	CHECKED *AWL*	SCALE *Not to scale*	UNITS *N/A*	SHEET *1 OF 1*
TITLE TOOL GRINDING			NUMBER F9	

Complete the tables with the correct information:

Machine guards and safety mechanisms used	How adjusted
1.	
2.	
3.	

Write here two checks for chisels before use to ensure they are safe:

1.

2.

Component dimensions	Angle:	Actual angle:	Error:
Scriber	*20° to 25°*		

Dot punch	*60°*		
Centre punch	*90°*		
Chisel	*60°, or other ...... (depending on the metal to be cut)*		

Name the coolant used while grinding the instruments:

Sketch:	*Safety screens*	*Machine work rest*
Sketch a detailed drawing of: 1. safety screens correctly adjusted for the offhand grinding machine. 2. machine work rest correctly adjusted for the offhand grinding machine.		

Test your newly sharpened tools and examine the point after checking.

1. Does it work properly?

2. Is the point still sharp?

List the safety precautions relevant to using grinding machines regarding:

1. Eye protection:

2. Machine guards:

Names of relevant laws specific to:

1. Grinding machines.

2. Abrasive wheels.

Start time and date:	End time and date:	Time taken (hours):

Witness testimony

I confirm that the tools were sharpened correctly without overheating the points, and that
. . . . used appropriate ppe and operated the pedestal grinding
machine safely.

Signed . . . . Job title . . . . Date . . . .

9 Machining engineering materials by turning

Exercise checklist

(Ask your assessor to initial the lower box for each exercise when you have completed it.)

Exercise no.	9.1	9.2	9.3	9.4	9.5	9.6	9.7
Initials							
Date							

The centre lathe

The machine shown below is a centre lathe. Centre lathes are versatile machines operated by a machine operator called a *turner*.

Centre lathes are used to turn cylindrical workpieces, the profiles of which include:

- parallel diameters
- tapered diameters
- drilled or reamed holes
- accurately bored holes
- externally threaded diameters
- internally threaded holes.

The work is held in a work-holding device and rotated. Cutting tools are secured in the machine's tool post and traversed along or across the workpiece's axis, thus removing material and producing the cylindrical shape.

Examples of typical turned components are:

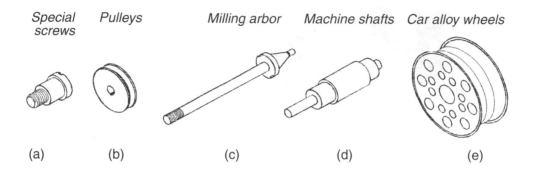

Special screws	Pulleys	Milling arbor	Machine shafts	Car alloy wheels
(a)	(b)	(c)	(d)	(e)

The fundamental parts of a centre lathe are labelled in the diagram below:

The centre lathe

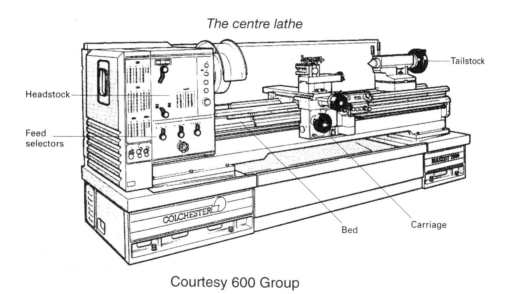

Courtesy 600 Group

- **Headstock** – The large cast iron housing on the left-hand side of a centre lathe. The headstock houses the machine's spindle and two gearboxes. The gearboxes enable the spindle's speed to be selected and the speed of the cutting tool movement to be adjusted, that is the feed rate. The headstock has various levers to set the speeds and feed rates.
- **Spindle** – Protruding from the headstock is the spindle's nose on which various work-holding devices can be attached. There are various methods of fixing the work-holding device (see page 288) to the spindle; two common types are illustrated below:

Camlock-type spindle nose

Screwed-type spindle nose

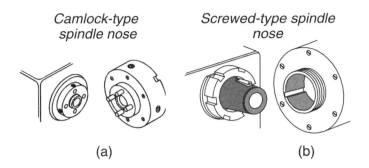

(a)

(b)

Section through a centre lathe's bed

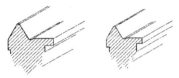

- **Bed** – Made of cast iron, the lathe bed is the machine's main slideway. Its top faces are accurately machined with flats and vee-shaped slides to provide location for the machine's carriage and tailstock. The carriage traverses along the outer slideways while the tailstock runs along the inner slideways. To resist twisting, the bed must be rigid and is therefore braced with ribs.
- **Carriage** – The centre lathe carriage can be traversed along the bed manually or automatically, carrying the cutting tool, the cross slide and the compound slide.

The carriage is an assembly of the following components:

A **The saddle** traverses along the bed's slideways. On top of the saddle are the cross slide, compound slide and the tool post. Usually, there is a nut on top of the saddle to enable it to be locked into position.

Detail of lathe carriage

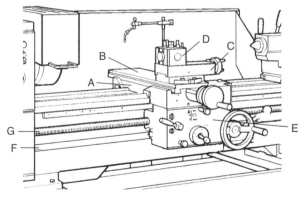

Courtesy The 600 Group

B **The cross slide** slides at 90° to the spindle's axis. The cross slide is operated by a graduated hand wheel feeding the cutting tool forward to remove material from the workpiece. Cross slides can be moved automatically under 'power feed' while facing the end of a bar.

C **The compound slide** can be set at any angle to allow the cutting tool to be fed at an angle to the workpiece, producing a taper. The compound slide is operated by a small hand wheel.

D **The tool post** houses the cutting tool.

E **The apron** is the front part of the carriage assembly. It houses various control levers and hand wheels. The apron is connected to the machine's gearbox via the feed shaft (F) and the lead screw (G).

- **Tailstock** – The tailstock is used for mounting various hole-cutting tools (e.g. drills and reamers) or a machine centre (to support a long workpiece). It is mounted on the bed's inner slideway and can be manually positioned. It can be locked into any position on the bed with a lever. The tailstock contains a barrel that can be wound forward and backwards by a hand wheel. The barrel can be locked into position with a second lever. There is a Morse-tapered hole in the tailstock barrel, and when measured from the bed, the height of the point of a machine centre mounted on the tailstock barrel is exactly the same as that of the machine centre mounted in the spindle.

A lathe tailstock

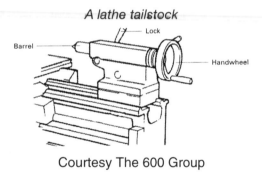

Courtesy The 600 Group

- **Centre lathe sizes** – The size of a centre lathe is measured in two ways:

  - The centre distance – the maximum distance between the tailstock and headstock centres.
  - The centre height – the maximum work radius that can be held.

Lathe measurements

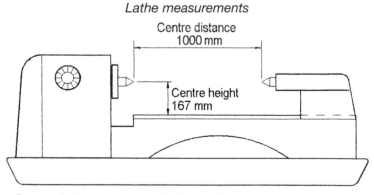

Centre lathe shown is 1000 mm 'between centres' and 167 mm 'centre height'

Mounting the work in a three-jaw self-centring chuck

Chucks are a quick and convenient method of work holding. Lathe chucks are mounted on the end of the spindle (the spindle nose). There are three main types of lathe chuck in common use:

- Three-jaw self-centring chuck.
- Four-jaw independent chuck.
- Spring collets.

All are described on page 288, together with other work holding devices.

For the first three exercises in this book, you will need to operate a centre lathe with a three-jaw self-centring chuck already fitted. Later, you will learn how to change work-holding devices.

The jaws of the three-jaw self-centring chuck are opened and closed by means of a chuck key inserted into one of the three square holes on the side of the chuck. When the chuck key is rotated, all the jaws open or close together so that the work can be held (approximately) centrally in the chuck. This type of chuck is only suitable for 'bright round bar' or 'bright hexagonal bar'.

A three-jaw self-centring chuck

Courtesy of Pratt Burnerd

Three-jaw chucks cannot be relied on for concentricity if the work-piece is removed and returned to the chuck, so do not remove the workpiece until it is finished.

Safety note! Always remove chuck key before starting machine.

Selection of spindle speeds

The spindle speed at which the work should be rotated depends on four factors; these are as follows:

- The diameter of the work being cut
- The type of material being cut
- The material from which the cutting tool is made
- Availability of coolant or cutting fluid

To calculate the correct spindle speed in rev/min for turning various materials, use the formula:

$$\text{rev/min} = \frac{1000 \times S}{\pi \times d}$$

where S = cutting speed of the work material*

d = work or drill diameter

π = 3.142 (often approximated to 3)

*See Appendix VIII for the cutting speeds recommended for various materials together with examples of spindle speed calculations.

Lathe cutting tool shapes

There are a number of different shapes of cutting tools suitable for use in a centre lathe. The cutting tool shape depends on the operation to be performed. Some of the most commonly used cutting tool shapes are shown and described below. All must be maintained with a keen cutting edge and resharpened or replaced if they become chipped.

- **Right-hand knife tool** – Used for machining light cuts on both faces or diameters. Although strength is limited, this tool shape is commonly used as it is versatile.

Right-hand knife tool

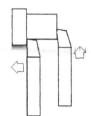

- **Left-hand knife tool** – Used for removing material from the headstock side of the work. Useful for making concentric workpiece faces and diameters as the workpiece need not be removed from the chuck.

Left-hand knife tool

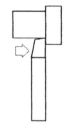

- **Chamfering tool** – Frequently used to cut chamfers on the corner of workpieces when the angle of the required chamfer is not critical. The chamfering tool shown is being used to cut a 45° chamfer.

Chamfering tool

- **Roughing tools** – Used for large amount of material removal. The angle of the front face on the tool reduces the force on the tool by making the chips (pieces of swarf) thinner. Force reduction allows faster feed rates to be used. As the workpiece's corners will not be square, it must be followed with a finishing cut, usually with a right-hand knife tool.

Straight nose roughing tool

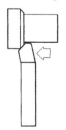

- **Parting off tool** – Used to separate the finished workpiece from the barstock. The parting off tool is slowly fed to the workpiece's centre line. Parting off is often a difficult operation; all moving parts should be clamped for maximum rigidity and you must take great care.

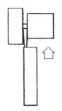

Parting off tool

- **Screw thread cutting tools** – The angle of the sharp point of screw thread cutting tools is equal to the thread angle, usually 60°. The feed rate for single start threads is one pitch of the thread per chuck revolution.

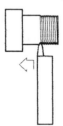

60° screw cutting tool

- **Facing tool** – Similar in appearance to the right-hand knife tool but the distinguishing feature is that the facing tool's point extends further to the left. As its name suggests, it is first choice for facing the ends of workpieces.

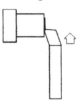

Facing tool

- **Grooving tools** – Produced with special profiles to make grooves in workpieces that can locate other components. They usually have short cutting projections to increase their strength. An example of using a grooving tool to cut an 'O' ring's groove is shown.

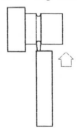

Grooving tool

- **Boring tools** – Available in a variety of shapes. The common feature of boring tools is a long shank that supports the cutting edge. The shank must be only as long as the hole is deep to reduce the effect of 'chatter' and vibration.

Square nose boring tool

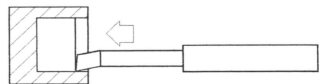

- **Radius tools** – Ground to suit the required radius of the corner of the workpiece. The drawings show the external radius tool cuts an external radius on the workpiece; the internal radius tool cuts an internal radius in the workpiece.

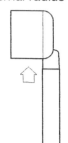

External radius tool

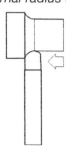

Internal radius tool

Machine centres

Machine centres are used on centre lathes for work holding, supporting long workpieces and for setting cutting tools at the correct height.

A machine centre is a hardened steel 60° point mounted on a Morse taper shank. The machine centre's point is located on the shank's axis. When fitted into a Morse-tapered hole, the point of the machine centre is on the centre line of the hole.

There are three types of machine centres in common use.

Full centre
General headstock and tailstock use

Half centre
For facing and small diameters

Rotating centre
For high speeds

Setting the cutting tool in the tool post

The centre lathe cutting tool must be mounted in the tool post. Its cutting edge is adjusted to match the height of the spindle's centre line; this is said to be 'on centre'.

Procedure for setting the cutting tool on centre

(a) Insert a machine centre into the tailstock barrel (its point is a guide when adjusting tool height).
(b) Clamp the cutting tool into the centre lathe's tool post with a minimum overhang (say 15 mm).
(c) Position the tool post so that the cutting tool's point faces the point of the machine centre.

Depending on the type of tool post fitted to your machine, you should then proceed by:

Either – Quick change 'camlock tool post'

(d) Slacken the grip of the hexagonal camlock screw.
(e) Turn the fine adjusting screw until the height of the cutting tool's point matches the height of the tailstock's centre.

Or – Four-way 'turret tool post'

(d) Insert or remove pieces of packing under the cutting tool to raise or lower its height. The height of the cutting tool's point should match the height of the tailstock's centre.
(e) Tighten cutting tool in tool post.

(f) Lock the hexagonal camlock screw.

(g) Check that height of the cutting tool's point still matches the height of the centre's point.

Quick to set cutting tool on centre, less rigid than four-way tool post

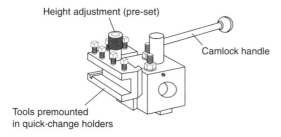

Height adjustment (pre-set)

Camlock handle

Tools premounted in quick-change holders

(f) Check that height of the cutting tool's point still matches the height of the centre's point.

More rigid than camlock type, slower to set cutting tool on centre

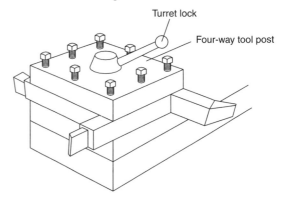

Turret lock

Four-way tool post

Selecting a feed rate

The feed rate is the distance the cutting tool advances along the workpiece's surface for each revolution of the spindle. The feed rate affects the quality of surface finish of the workpiece: a fine feed (say 0.04 mm/rev) would give a better finish than a coarse feed (say 0.5 mm/rev). A coarse feed would remove material faster.

Usually a chart giving various feed rates for certain lever positions is on the headstock. The feed rate is set by adjusting the position of the levers. Use coarse feeds for roughing out workpieces and fine feeds for finishing accurately to size.

The shape of the tool nose also affects the surface finish, see figure.

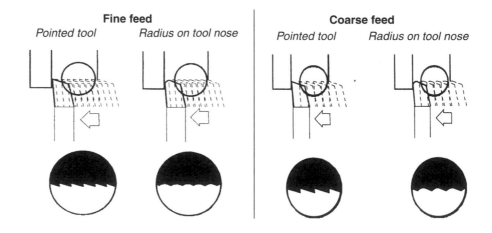

Fine feed

Pointed tool   *Radius on tool nose*

Coarse feed

Pointed tool   *Radius on tool nose*

Depth of cut

Removal of material by turning the depth of cut selected is dependent on a number of factors that must be taken into account, these include:

• type of cutting tool material
• cutting tool condition and geometry

- type of material being cut
- availability and type of coolant
- feed rate
- power of the machine's motor
- condition of the machine
- rigidity of the machine's tool post and work-holding device.

If vibration starts, stop the machine immediately and check if the tool is sharp and reduce the depth of cut and/or the feed rate.

Using coolant

Both coolants and cutting fluids are liquids used on centre lathes and other machine tools for:

- cooling the workpiece and the cutting tool
- lubricating the cutting process
- washing away pieces of swarf
- improving the surface finish of the workpiece
- preventing workpiece from corrosion

Coolant – Coolants are stored in a tank under the centre lathe and pumped out through a small pipe. The coolant can be directed to the point at which the actual metal cutting takes place. A common type of general-purpose coolant is known as 'soluble oil'. Soluble oil is a thick oil mixed with water to form a milky coloured solution commonly called '*suds*'.

Cutting fluid – For heavy-duty work or for particularly long production runs, cutting fluids that have better lubricating properties than coolants are used. Mineral oils and synthetic oils are common cutting fluids. Cutting fluids are more expensive and efficient than soluble oil (coolant). Various blends of oils are suited to different cutting conditions. A cutting fluid supplier may recommend a particular cutting fluid for a given application.

Positioning of centre lathe guards

Turning can be a hazardous process. You must never touch moving workpieces or attempt to clear away swarf while the centre lathe is running. It is essential that guards are positioned and used appropriately when the machine is working to offer maximum protection for you – the operator.

Centre lathe chuck guard

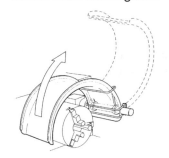

Courtesy of Nelsa Machine Guarding Systems Ltd.

- **Centre lathe chuck guard** – A popular type of chuck guard for centre lathes. The guard may be transparent so that the workpiece can be seen. In the closed position, it protects the operator from flying swarf and from being splashed by coolant. When hinged open, there is easy access to the work area.
- **Sliding lathe guard** – A larger type of guard that is very efficient, protecting the turner from contacting the chuck and long workpieces as it guards the whole of the cutting area. It is simple to slide into position and is transparent. Electrical interlocks are sometimes fitted to sliding lathe guards to prevent machines from being started before the guard has been correctly positioned.
- **Bellows guard** – A bellows cover can be attached to the lathe to guard the lead screw and the feed shaft. This type of guard offers protection to the operator by preventing loose clothing from entanglement and protects the machine by keeping abrasive swarf away from precision components.

Sliding lathe guard

Bellows guard covering the lead screw and feed shaft

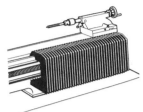

Courtesy of Nelsa Machine Guarding Systems Ltd.

Courtesy of Silvaflame Co. Ltd.

The Health and Safety at Work Act 1974 states that you must take reasonable care of the health and safety of yourself and other people who may be affected by your acts. This means you must **always** use guards provided and report any operating failures or damaged guards to your supervisor.

Operating the centre lathe

Never **work on any machine until you have been shown how to operate it properly by a suitably qualified person.**

Your supervisor will arrange for you to be shown how to operate a centre lathe and introduce you to the following features:

- Power supply and how to isolate it
- The various guards on the machine
- Spindle stop, start and reverse controls
- Hand feed dials on the hand wheel for cross feed
- Hand wheels for:
  - cross feed
  - longitudinal feed.
- Power feed control for:
  - traversing along the work
  - surfacing across the work.
- Spindle speed selection levers
- Feed selection levers
- Coolant/cutting fluid supply
- Clamps for the tailstock and the saddle
- How to manage 'backlash'
- Emergency stop procedure.

If you do not understand what you have been shown or if you are unsure how to proceed, you must *ask!*

Never operate a machine until the slideways are clear and all tools and equipment are away from moving machine parts.

Never leave a machine unattended while in motion.

When you have practised working the centre lathe using pieces of spare material and have become confident about the controls, ask your supervisor for permission to start the turning exercise.

EXERCISE 9.1   Turning exercise

Make the turning exercise using the procedure and tools indicated in the planning section below. Your finished work should be within the stated drawing limits.

On completion of the task, carefully measure your work with rules for lengths and micrometer for diameters and write in the table the actual sizes, noting any errors.

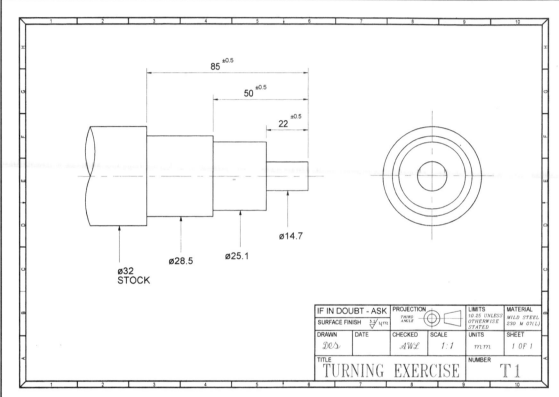

IF IN DOUBT - ASK	PROJECTION		LIMITS	MATERIAL
	THIRD ANGLE		±0.25 UNLESS OTHERWISE STATED	MILD STEEL 230 M 07(L)
SURFACE FINISH 3.2 µm				
DRAWN *DCS*	DATE	CHECKED *AWL*	SCALE *1:1*	UNITS *mm*	SHEET *1 OF 1*
TITLE				NUMBER
TURNING EXERCISE				T1

Drawing dimensions: 85 ±0.5, 50 ±0.5, 22 ±0.5, ø14.7, ø25.1, ø28.5, ø32 STOCK

Planning:
Complete the tables with the correct information for your centre lathe:

Procedure	*Tools and equipment selected*
1. Acquire material and check for size. 2. Hold work in a three-jaw chuck with 100 mm protruding. 3. Set spindle speed. 4. Set cutting tool at machine's centre height. 5. Face off one end of the bar. 6. Turn each diameter, starting with the biggest. 7. Deburr all sharp edges. 8. Stamp your name at the end of your work. 9. Isolate your machine from the electricity supply. 10. Clean the work area and machine.	*Metric 300-mm rule* *Centre lathe* *Right-hand knife tool* *0–25-mm micrometer* *25–50-mm micrometer* *Centre (rotating or dead)* *½ lb. hammer* *Letter stamps*

Make of machine used:	One graduation on the cross slide reduces the work's diameter by:
Centre height	Centre height of the biggest centre lathe in your workshop
Distance between centres	Distance between centres on the smallest centre lathe in your workshop
Model of machine	Quantity of spindle speeds available on the machine:

Calculated spindle speed for turning ø30-mm mild steel with high-speed steel (HSS) tooling:	
Calculated spindle speed for turning ø100-mm mild steel with HSS tooling:	

Write here the emergency stop system on the machine and state how it can be checked:

Write an account of when you needed to ask your supervisor for help or advice and the actions that were agreed.	
Draw a sketch showing how the height of the cutting tool is adjusted to match the workpiece's (spindle's) centre line.	

Finished inspection report:

Component dimensions (mm)	Limits:	Actual size:	Error:
Length 22	22.5 21.5		
ø28.5	28.75 28.25		
ø25.1	25.35 24.85		
ø14.7	14.95 14.45		

Results:

Is the workpiece to the drawing specification?

YES	NO

If completed workpiece is below drawing specification, list here the reasons for any errors, stating clearly how the errors will be avoided in future.

Start time and date:	End time and date:	Time taken (hours):

Witness testimony

I confirm that the turning exercise was made within the stated limits and that. . . . . .carried out the work in a safe manner using appropriate roughing and finishing cuts and clearing the work area on completion of the task.
The problems outlined by the trainee above were discussed with me and the actions agreed were carried out in a safe manner.

Signed . . . . .Job title . . . . .Date . . . .

Thread cutting using dies

A simple and efficient method of cutting small external threads on a centre lathe is to use a split button die in a diestock. The split button die is mounted and set in the diestock as described in Chapter 1, Basic Bench Fitting. The front face of the tailstock barrel can be used to ensure that the diestock is square to the workpiece.

Cutting external threads on a centre lathe using dies

The procedure is as follows:

(a) Ensure that the outside diameter of the material to be threaded is the same as the thread's major (outside) diameter, for example M10 = ø10 mm.

(b) Lightly chamfer the material's front face by setting the chamfering tool at the required angle as shown. The chamfer provides a lead that helps the die to start cutting the thread.

Setting a tool for chamfering

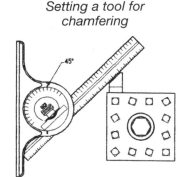

Chamfering the workpiece

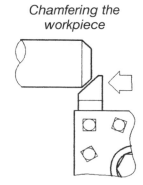

Set-up ready to thread while holding the die square to the work

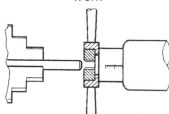

(c) Isolate centre lathe's electricity supply.

(d) Remove the cutting tool.

(e) Disengage the gearbox so that the spindle is free to turn.

(f) Set the diestock to cut at its maximum diameter (refer to Fitting section, pp. 232–33).

(g) Turn the tool post so that one of its plane surfaces faces the workpiece.

(h) Bring the tailstock up to the end of the workpiece and lock it to the bed with its front face about 20 mm away from the work.

(i) Position the diestock in the gap between the tailstock barrel and the workpiece.

(j) Move the barrel forward so that it **lightly** traps the diestock against the workpiece.

(k) Turn the handle of the diestock so that it rests on the tool post.

(l) Put a small amount of cutting oil onto the workpiece.

(m) Cut the thread by rotating the spindle with your left hand and gently move the tailstock barrel forward with your right hand keeping the diestock square to the workpiece.

(n) Move the tailstock barrel back each time you relieve the die of swarf (each half turn of the chuck).

(o) Continue thread cutting until the correct length is threaded.

(p) Check the thread with thread gauges or the mating component, if it is available.

(q) If necessary, the diestock can be reset with a screwdriver to finish the thread's profile (see page 229).

EXERCISE 9.2   Clamp studs

Make the clamp studs using the procedure and tools indicated. Your finished work should be within the drawing limits.

On completion of the task, carefully measure your work and write in the table the actual sizes, noting any errors.

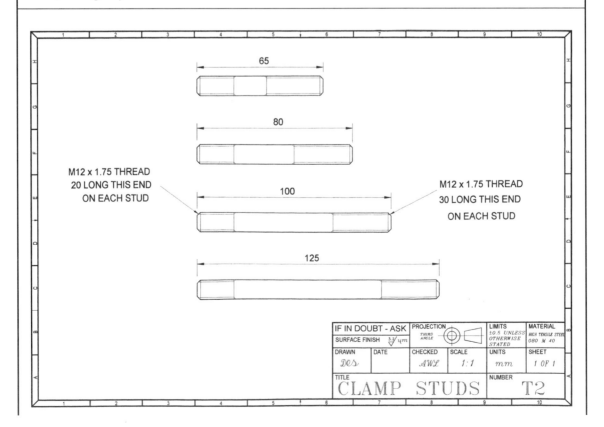

65

80

M12 x 1.75 THREAD
20 LONG THIS END
ON EACH STUD

100

M12 x 1.75 THREAD
30 LONG THIS END
ON EACH STUD

125

IF IN DOUBT - ASK	PROJECTION THIRD ANGLE		LIMITS ±0.5 UNLESS OTHERWISE STATED	MATERIAL HIGH TENSILE STEEL 080 M 40
SURFACE FINISH $3.2/\mu m$				
DRAWN *DES*	DATE	CHECKED *AWL*	SCALE *1:1*	UNITS *mm*	SHEET *1 OF 1*
TITLE CLAMP STUDS				NUMBER T2

Planning:

Procedure	Tools and equipment selected
1. Acquire material and check for size. 2. Saw length into pieces of approximate length. 3. Set spindle speed. 4. Set cutting tool at machine's centre height. 5. Face off ends to length. 6. Chamfer the corners of the work with a chamfering tool. 7. Isolate your machine from the electricity supply. 8. Cut the thread to length with the stock and die. 9. Check the threads with a gauge. 10. Clean the work area and machine.	*Metric 300-mm rule* *Hack saw.* *Centre lathe* *Right-hand knife tool* *Centre (rotating or dead)* *M12 × 1.75 split button die* *Diestock* *Screwdriver* *Thread gauge*

Research:

Facilities available for clearing swarf:	
Correct safety clothing to be used while clearing swarf:	
Where swarf should be put prior to from works:	

List below six items of safety clothing to be used while doing the exercise:

1.

2.

3.

4.

5.   6.

| Write the name of the cutting fluids used: | 1. when turning |
| | 2. when cutting the threads |

Finished inspection report:

Component dimensions (mm)	Limits:	Actual size:	Error:
M12 × 1.75 thread form	*To fit M12 'thread gauge'*		
Length of stud 1 65	*65.5* *64.5*		
Length of stud 2 80	*80.5* *79.5*		
Length of stud 3 100	*100.5* *99.5*		
Length of stud 4 125	*125.5* *124.5*		

Results:

Are the workpieces within the drawing specification?   YES   NO

If the completed workpiece is below drawing specification (for example, is the thread correctly formed and free from tears? Is the thread square to the axis of the workpiece?), list the reasons for any errors, stating clearly how the errors will be avoided in future.

Start time and date:	End time and date:	Time taken (hours):

Witness testimony

I confirm that the set of clamp studs was made within the stated limits and
that. . . . . carried out the work in a safe manner. The work area was
cleared and the swarf was removed to a safe place on completion of the task.

Signed . . . . Job title . . . . Date . . . .

Facing to length

When the overall length of a workpiece is specified with close limits,
it is not possible to cut the length within the limits by using the
scale on the longitudinal traverse hand wheel. There are often no
graduations on the longitudinal hand wheel, and, when available, they
are quite coarse.

A reliable way to turn accurate lengths is as follows:

(a) Check if the compound slide is parallel (set at $0°$) to the lathe bed.
(b) Face both ends roughly to the correct length.
(c) Remove the workpiece from the chuck and accurately measure it
    with a micrometer.
(d) Calculate the difference between the actual size and the
    required size.
(e) Put the workpiece back in the chuck and secure it.
(f) Set the compound slide's reading to zero by turning the scale and
    not moving the slide.
(g) Turn the longitudinal hand wheel until the tool **just** touches the
    end of the workpiece.
(h) Lock the carriage to the bed in this position with the locking
    screw/bolt.
(i) Feed the cutting tool point towards the chuck the calculated
    amount.
(j) Face the end off.
(k) The workpiece should now be of the required length.

Chamfering

When cutting small chamfers on a lathe, a quick method is to use a
chamfering tool. The angle of the front face of the cutting tool is set
by turning the tool post so that the cutting tool's front face matches
the blade of a protractor (see figure 'Setting a tool for chamfering'
on p. 277).

With the workpiece secured in the machine, carefully feed the
chamfering tool forward until the chamfer is formed to the specified
length.

Lathe tool materials

Nearly all centre lathe cutting tools are made from either HSS or
tungsten carbide. HSS is cheaper and tougher (shock resistant) than
tungsten carbide and can be more easily shaped by grinding. HSS is
frequently used in the manufacture of small quantities of components
and development work. The HSS cutting edge is fastened to a mild
steel shank, usually by welding or by clamping it into a special tool
holder.

Tungsten carbide is harder, more brittle and longer lasting than HSS, allowing faster spindle speeds to be used. Tungsten carbide is frequently chosen for large batch production. The cutting tip, which is made from tungsten carbide, is either clamped or permanently brazed to the tool's shank.

Clamped tungsten carbide tool
Very quick to change tip for a replacement

Brazed tungsten carbide tool
Brazed joint increases the toughness of the tool

Welded HSS tool
HSS is an extremely tough cutting material

Clamped HSS tool
Economical cutting tool material

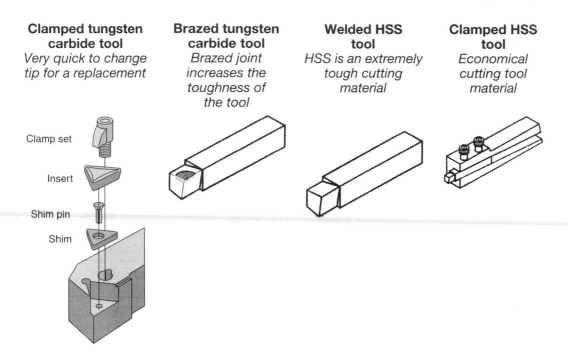

Clamp set

Insert

Shim pin

Shim

Angles on turning tools

Lathe cutting tools must be kept sharp or the surface finish of the work will be poor. When using a clamped tungsten carbide tip tool, if the tool tip becomes worn or chipped, a new tip edge can easily be fitted by unscrewing the old tip and replacing it or turning it round. Brazed tungsten tools and all types of HSS tools must be resharpened. Your supervisor will usually do this for you. All lathe cutting tools are ground in such a way as to avoid any part of the tool, except the actual cutting point, touching the workpiece. Consider the right-hand knife tool below; the four important angles are:

Right-hand knife turning tool

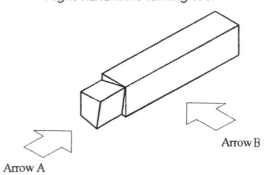

Arrow B

Arrow A

Right-hand knife tool

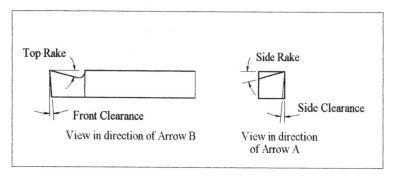

View in direction of Arrow B

View in direction of Arrow A

Top rake	**Front clearance**	**Side rake**	**Side clearance**
(15° for mild steel) Angle of slope towards operator on top face of cutting tool.	(8° for most materials) Angle between front face of cutting tool and workpiece.	(15° for mild steel) Angle of slope towards tailstock on top face of cutting tool.	(6–8° for most materials) Angle between side face of cutting tool and workpiece.

Note. Rake angles should be increased for soft materials (e.g. aluminium) and decreased for hard materials (e.g. high-carbon steel).

Mounting cutting tools in the tailstock

The Morse taper bore in the tailstock barrel can be used to mount the following:

- Machine centres, either rotating or 'dead'
- Drill chucks that can be used for holding centre drills, small parallel shank drills and reamers
- Drills with Morse taper shanks mounted directly into the tailstock barrel
- Machine reamers of various sizes
- Tapping heads.

Once the workpiece is secured in the chuck, check that the Morse-tapered bore of the tailstock is clean and free from foreign matter. Wipe the outside of the drill's shank and push it firmly into the tailstock. If the tailstock Morse taper does not match that of the drill, a Morse taper converting sleeve must also be used.

When a drilled or bored hole is required:

1. Start by centre-drilling a small dimple at the end of the bar stock.

   (a) Mount a drill chuck in the tailstock
   (b) Fit a centre drill and set the machine spindle speed (see Appendix VII)
   (c) Lock the tailstock to the bed
   (d) Carefully feed the barrel forward so that the centre drill cuts into the end of the workpiece

Centre drilling

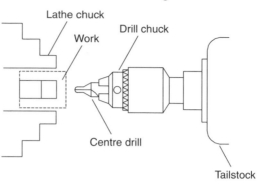

(e) The correct depth for a centre-drilled hole is about three-quarters of the way along centre drill's 60° point.

Centre depth correct

Centre drill not deep enough

Centre drill too deep

Fitting the drill

2. To drill an accurately sized hole:

(a) Select a drill to pilot drill (predrill) the workpiece with a smaller sized drill than the finished drill size. This will minimise the chances of the hole being oversized.
(b) Mount the pilot drill in the tailstock securely.
(c) Wind the tailstock handle until the measuring scale on the barrel reads zero.
(d) Move the tailstock until the drill point touches the workpiece and lock the tailstock to the bed.
(e) Start the centre lathe's spindle at the correct speed.
(f) Apply coolant or cutting fluid.
(g) Wind the drill forward until the correct depth is reached.
(h) Retract the drill from the newly drilled hole.
(i) Stop the spindle and remove the pilot drill.
(j) Fit the final size drill or reamer to finish the hole accurately to size (as stages b–i above).

Drilling to depth

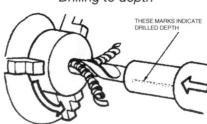

THESE MARKS INDICATE DRILLED DEPTH

EXERCISE 9.3   Drill stand legs

Make the drill stand legs using the procedure and tools indicated. Your finished work should be within the drawing limits.

On completion of the task, carefully measure your work and write in the table the actual sizes, noting any errors.

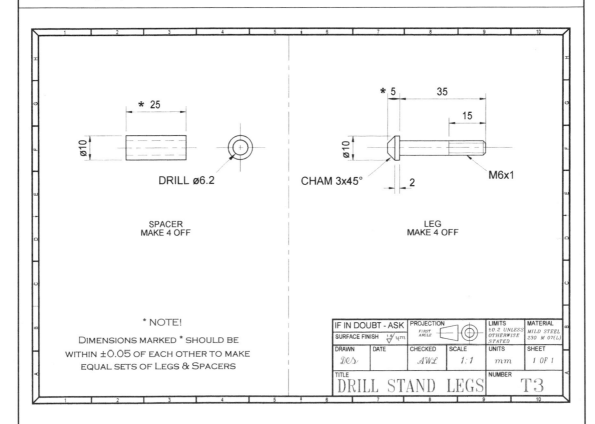

Planning:

Procedure	*Tools and equipment selected*

Spacers
1. Acquire material and check for size.
2. Saw four pieces for spacers 26 mm long at a bench vice.
3. Face both ends of each spacer to length within ±0.05 mm.
4. Centre drill and drill through each spacer in turn.

Legs
1. Acquire material and check for size.
2. Saw into two pieces, each approximately 90 mm long.
3. Turn ø6 × 35 mm long on each end of material.
4. Cut M6 × 1 thread, 15 mm long on each end of each piece.

Tools and equipment selected

Hacksaw.
Metric 300-mm rule.
Metric 0–25 micrometer.
Centre drill.
Drill chuck.
Chamfering tool.
Ø6.2 drill.
Right-hand knife tool.
M6 × 1 Stock & die

Procedure (cont.)	*Tools and equipment selected (cont.)*
5. Saw each piece in half at a bench vice. 6. Turn ends to 5 mm ±0.05 mm shoulder. 7. Cut chamfers on legs with a chamfering tool. 8. Remove all burrs from the workpieces. 9. Isolate machine and clean the work area.	

Write three reasons why it is important that the area around the centre lathe is kept clean and tidy:

1.

2.

3.

Sketch:

Make two sketches showing how the turning tools were held

1. in the tool post
2. in the tailstock

Tool post
Tailstock

Finished inspection report:

Component dimensions (mm)	Limits:	Actual size:	Error:
1. Spacer lengths:	*Set to be ±0.05 of each other*		
2. Foot lengths:	*Set to be ±0.05 of each other*		
3. Plain lengths:	*35.2* *34.8*		
4. Hole diameter:	*6.4* *6.0*		

Results:

Is the workpiece to the drawing specification?

YES	NO

If completed workpiece is below drawing specification, write here the reasons for any errors, stating clearly how the errors will be avoided in future.

Start time and date:	End time and date:	Time taken (hours):

Witness testimony

I confirm that the drill stand legs have been made within the stated limits, managing backlash in the machine slides.
. . . . carried out the work in a safe manner and cleared the work area on completion of the task.

Signed . . . . Job title . . . . Date . . . .

Work holding devices

There are five different types of work-holding devices commonly used on centre lathes:

1. Three-jaw self-centring chuck
2. Four-jaw independent chuck
3. Face plate
4. Catch plate (for turning between centres)
5. Spring collets

The applications and uses for these work-holding devices are described below:

1. **Three-jaw self-centring chuck** – The most widely used chuck on a centre lathe. It can be used only to hold bright round barstock, bright hexagonal barstock or previously turned components.

   Insert the chuck key in the square and turn it clockwise; all three jaws move in together to grip the workpiece. Female jaws can be used to hold large diameter workpieces.

   The drawback with this type of chuck is that if a workpiece is removed and replaced back into the chuck, the workpiece no longer runs exactly true (concentric). There is no means to correct this run-out.

   When changing the jaws, ensure that the set of jaws you are going to fit are an actual set (they are usually numbered with a code). The slots on the front face of the chuck and each jaw are numbered 1, 2 and 3. Make sure that jaw No. 1 goes into slot No. 1 and is 'picked up' by the scroll first, then insert jaw No. 2 in slot No. 2 and so on. **Never leave a chuck key in a chuck when not in use. If the machine is started with the chuck key in, the chuck key will fly out causing injury or damage.**

2. **Four-jaw independent chuck** – The jaws of this type of chuck are each adjusted independently to grip the workpiece. Workpieces of any constant cross-sectional shape (e.g. square, octagon or black round bar) can be accurately gripped if it is set up with the aid of a dial test indicator.

   Four-jaw chucks are used for holding part finished work that must run absolutely true; however, they are slow to set up.

3. **Face plate** – The face plate is used for holding large irregular-shaped workpieces. The workpiece is bolted to the face plate through its slots. Sometimes, counter balance weights are necessary to ensure the smooth running of the machine. They must be **securely** fixed to the face plate for safety. Face plates are only used for short workpieces and are very slow to set up.

4. **Catch plate** – The catch plate has a peg or slot that transmits the rotary drive to the workpiece via a carrier when it is held between centres as shown.

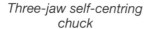

Three-jaw self-centring chuck

Courtesy of Pratt Burnerd

Four-jaw independent chuck

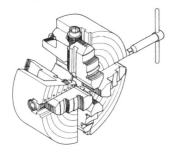

Face plate

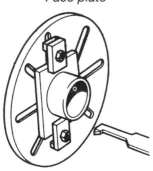

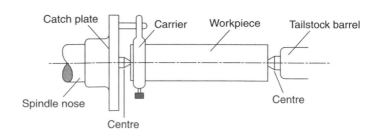

Spring collet

Work is normally held in this way when all the diameters must be concentric and it is to be machined all over.

5. **Spring collets** – Spring collets can be used for holding work-pieces when a higher degree of concentricity is required than that provided by a three-jaw chuck. They also cause little damage to the workpiece's surface.

Spring collets are special inserts that may be inserted into the centre lathe's spindle. When the operator pulls a lever (or turns a hand wheel), the spring collet is drawn back into the taper of the lathe's spindle, closing up and gripping the workpiece.

Changing the work-holding device

As introduced at the start of this unit, there are a number of different methods of attaching the work-holding device to the machine's spindle. Each machine tool manufacturer has slightly different designs of spindle, the two most common methods being the screwed type and the camlock type.

The work-holding device can be changed quite quickly and easily. Your supervisor will show you the correct method of changing them on your type of machine.

Be sure to:

Correct lifting technique
Bend at the knees and keep your head well up

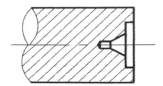

- isolate the machine at the mains before starting
- use correct lifting technique when handling heavy chucks
- adequately protect the machine bed from damage by use of a piece of wood called a *chuck board* or a similar device
- replace chucks to the appropriate storage rack, keeping them away from gangways or emergency exits
- thoroughly clean the spindle nose and mating faces on the chuck before assembly
- tighten work-holding device securely before starting the machine
- check if the chuck runs true on both its front face and its periphery (outside edge) before turning the centre lathe on.

Turning between centres

When a workpiece is specified as requiring concentric diameters and it cannot be completed in one setting in a lathe chuck, the most efficient method of turning it is to hold it between the lathe centres.

Procedure for turning between centres

A protected centre-drilled hole

(a) Centre drill the workpiece at both ends, that is, on a centre lathe using a three-jaw chuck or in a drilling machine. The centre drilled holes form a 60° surface for the centres to locate on. It is good practice to protect the centre with a small recess.

(b) Mount the centre lathe's catch plate to the spindle nose.

(c) Clean a centre and insert it into the Morse-tapered hole in the spindle nose (a special bush is often required to convert Morse taper sizes). This centre is called a 'live centre' because it rotates with the spindle. The point of the centre is now exactly in line with the spindle's axis.

(d) Clean and insert a centre into the tailstock barrel.

Mounting the headstock centre

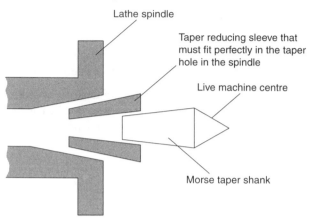

Lathe spindle

Taper reducing sleeve that must fit perfectly in the taper hole in the spindle

Live machine centre

Morse taper shank

Checking centre alignment

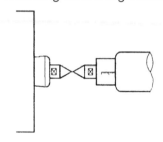

(e) Move the two centres together so that their points nearly touch. If there is an alignment error (i.e. if points do not match), see your supervisor; the tailstock may need adjustment.

(f) Select a lathe carrier that fits the workpiece's outside diameter and connects to receive the drive from the catch plate.

There are two types of carrier:

(a) Cranked tail carriers for slotted catch plates
(b) Straight tail carriers for catch plates with driving pegs.

Two types of lathe carrier

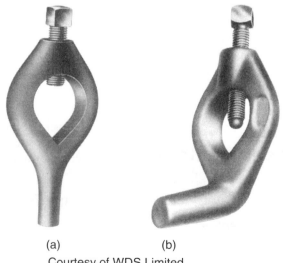

(a)   (b)
Courtesy of WDS Limited

(g) Attach the carrier to the headstock end of the workpiece.
(h) Clean the centre points and centre holes in the workpiece.
(i) Set the tailstock barrel with about 60 mm to 80 mm protruding from the tailstock body.
(j) Lock the tailstock to the bed in a position allowing a gap between the centres slightly greater than the length of the workpiece.

(k) Put the left-hand end of the workpiece in the headstock centre and wind the tailstock barrel forward until the tailstock centre supports the right-hand end of the workpiece.

(l) Check that the carrier is receiving drive from the catch plate.

(m) Lock the carrier into position on the workpiece (the workpiece should be able to rock a little).

(n) Check that the workpiece has no end float.

(o) Lock the tailstock barrel in position.

Workpiece mounted between centres

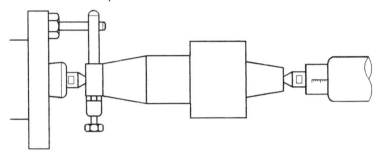

Taper turning using the compound slide

The compound slide of a centre lathe can be swivelled and fixed at any angle to the spindle. When the compound slide is set at an angle, the cutting tool can be manually wound forward along the angular compound slide generating a taper on the workpiece. Offsetting the compound slide is a simple and effective method of turning tapers that are too long to be turned with a form tool. A good finish is only achieved by carefully winding the cutting tool forward, slowly and steadily.

The length of taper turned by this method is restricted to the length of the compound slide. Most centre lathes' compound slides are limited to about 125-mm long, so this would be the maximum length of taper that could be turned.

Procedure for taper turning using the compound slide

(a) Look at the drawing and notice if the angle on the workpiece is indicated as the included angle or the angle per side.

(b) If the angle indicated on the drawing is the 'included angle', it must be halved to find the angle per side.

(c) Set the cutting tool on centre height.

(d) Turn the compound slide to half of the included angle, that is the angle per side.

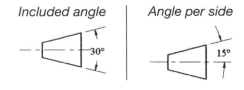

Included angle | *Angle per side*

Setting the compound slide angle

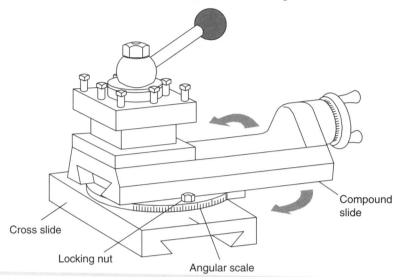

Cross slide

Locking nut

Angular scale

Compound
slide

(e) Wind the compound slide out as far as it will go.
(f) Turn the tool post so that the tool is straight.
(g) Move the tool with the cross slide so that it is near the end of
the workpiece.
(h) Lock the saddle to the bed.
(i) Start the machine's spindle.
(j) Apply a 'cut' with the cross slide hand wheel.
(k) Feed the tool forward with the compound slide hand wheel
slowly and steadily.
(l) When the tool stops cutting, wind the compound slide back to
the start.
(m) Repeat stages j to l until the taper is as required.

Taper turning using the compound slide

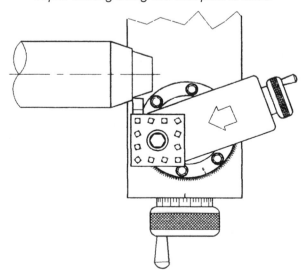

EXERCISE 9.4 Alignment tool

Make the alignment tool using non-ferrous material. Make a planned operation sequence and list of tools required and get it approved by your supervisor before starting work.

Your finished work should be within the drawing limits. On completion of the task, carefully measure your work and write in the table the actual sizes, noting any errors.

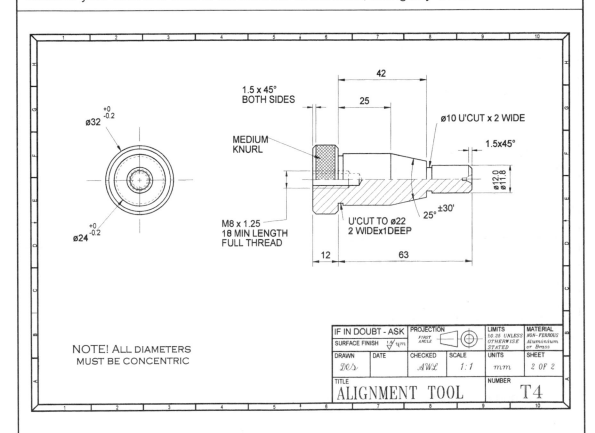

Planning:

Procedure	Tools and equipment selected
1. Acquire material and check for size.	
2.	
3.	
4.	
5.	
6.	
7.	

Procedure (cont.)	*Tools and equipment selected (cont.)*
8.	
9.	
10.	

Research:

What safety checks were done on the tools before starting work:	
What safety checks were done on the equipment before starting work:	
What checks were done on the materials before starting work:	
State any special disposing procedure for non-ferrous material:	

Write a short statement in answer to the following questions:

1. Why 'turning between centres' is the best and most suitable work-holding method for this job.

2. How the centres were protected from overheating?

3. What would go wrong if the centres were overheated?

Sketch:

Make a sketch showing the set-up for turning between centres.

Finished inspection report:

Component dimensions (mm)		Limits:	Actual size:	Error:
ø12		12.0 mm 11.8 mm		
ø24		24.0 mm 23.8 mm		
Root diameter of undercuts	ø10	9.75 mm 10.25 mm		
	ø22	21.75 mm 22.25 mm		
25° angle		25° 30' 24° 30'		
Length 12		12.25 mm 11.75 mm		

Results:

Is the workpiece to the drawing specification?

| YES | NO |

If completed workpiece is below drawing specification, list here the reasons for any errors, stating clearly how the errors will be avoided in future.

| Start time and date: | End time and date: | Time taken (hours): |

Witness testimony

I confirm that the alignment tool was planned correctly and made within the stated limits.
. . . . .carried out the work in a safe manner and cleared the work area on completion of the task.

Signed . . . . .Job title . . . . Date . . . .

Boring on a centre lathe

A process known as boring is carried out on a centre lathe when an existing hole is to be accurately finished to a specific size. The technique is particularly useful for machining unusual sized holes if a reamer is not available or finishing internal diameters when concentricity in essential.

Components being bored cannot be mounted between centres, but any of the other work-holding devices described on page 288–89 would be suitable. The cutting tool used for boring must be selected as follows:

1. **Types of boring tool in common use:**

   (a) **Butt-welded HSS boring tools** – Useful for a variety of tool room and specialist tasks, the shape of the cutting edge can be ground for individual requirements.

   (b) **Carbide-tipped boring tools** – The carbide tip is secured in the tool holder by a small screw. The tip's hardness allows fast cutting speeds to be used. The tool's cutting angles cannot be ground to suit particular cutting conditions; consequently, carbide-tipped boring tools are used for batch production.

Butt-welded, HSS boring tool

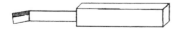

Carbide-tipped boring tool

2. **Boring tool length** – The length of tool overhang must be kept to a minimum so that there is the least chance of setting up any vibration and 'chatter'.

Setting a boring tool with minimum overhang

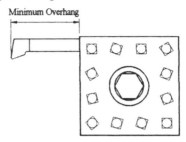

Boring tool viewed 'in situ' through the machine spindle

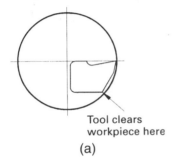

Tool clears
workpiece here

(a)

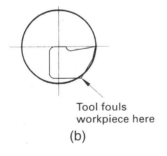

Tool fouls
workpiece here

(b)

3. **Boring tool clearance** – Tool angles, particularly clearance angle, must be carefully checked before starting to cut metal.

Examine the two figures. Although the boring tool used is the same in both the drawings, the workpiece is smaller on the bottom drawing. Notice that the heel of the boring tool fouls (catches) the smaller workpiece on the bottom drawing. If this is the case, the corner of the tool is removed by grinding. On the larger bore, boring tool does not foul the workpiece at all. A good position to see if a boring tool is fouling a workpiece is to look through the spindle's bore and view the boring tool inside the workpiece.

Procedure for boring

(a) The workpiece is mounted in a suitable work-holding device, centre-drilled and drilled to 'rough out' the hole. Usually this roughing out takes the bore within about 3 mm of the finished size.
(b) Mount the boring tool in the tool post on centre height.
(c) Position the lathe's saddle and cross slide so that the boring tool point is touching the near corner of the workpiece.

Setting centre height

(d) Check if there is clearance under the cutting edge.
(e) Calculate and set the correct rev/min for the spindle (use the finished bore diameter as the work diameter).

$$\text{rev/min} = \frac{1000 \times S}{\pi \times d}$$

(f) Wind the cross slide towards you so that it is ready to take a cut.
(g) Start the machine's spindle.
(h) Feed the boring tool into the workpiece. If the bore goes completely through the workpiece, power feed can be used (see note i below).

Boring on a centre lathe

Courtesy of Silvaflame

(i) Wind the saddle back so that the boring tool is clear of the work-piece.
(j) Measure the bore size using either:
  • internal micrometers (for large bores)
  • telescopic gauges (for small bores).

Measuring bore with an internal micrometer

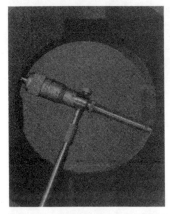

Courtesy of Moore and Wright

Measuring bore with 'telescopic bore gauges'

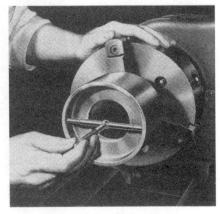

Courtesy of Starrett Ltd.

(k) Apply roughing and finishing cuts with the cross slide hand wheel to take roughing and finishing cuts until you have completed the boring operation.

Note.

1. The use of stops will assist boring features to depth and turning accurate lengths. Ask your supervisor if stops are available for the bed of your centre lathe. He will show you how to fit and use them.

A saddle stop is fixed to the lathe bed to set traverse lengths for turning operations

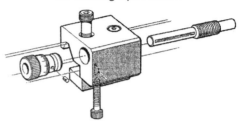

Courtesy of The 600 Group

2. If stops are not available, and the depth of a bore is specified, the saddle should be locked to the bed. The depth of bore can then be determined by winding the compound slide forward while reading its dial.
3. The boring tool sometimes deflects slightly during cutting. When this happens, the bore in workpiece is left slightly undersize. If this is happening and your workpiece is near to the finished size, repeat the pass at the same cross slide setting. The boring tool will not deflect on the second pass, so the size should now be as expected.

The mandrel

A mandrel is a round bar that is slightly tapered. It is pressed into a hole in a workpiece and mounted between centres on a lathe. The workpiece can then be machined on its outside surfaces in such a way that the outside diameter is concentric with the inside diameter. Mandrels are used when an outside diameter is to be turned concentric to its bore, for example, when turning a pulley.

The key features of a mandrel are as follows:

- centre holes at each end are large and recessed
- ground surface finish machined from the centre holes
- slight taper of about 0.5 mm per metre (0.0005″ taper per inch)
- usually a '+' sign is etched on mandrels at the bigger end.

Mounting workpiece on a mandrel

(a) Select a mandrel to fit the bored (or reamed) workpiece.
(b) Remove all burrs from the workpiece.
(c) Insert the small end of the mandrel into the workpiece's bore.

(d) Use a mandrel press to firmly push the mandrel into the work-piece's bore.

When turning with a mandrel, mount the mandrel with its large end (marked +) at the headstock end of the centre lathe (this tends to keep the workpiece tight).

The turning process is exactly the same as turning between centres, the carrier is attached to a flat on the end of the mandrel. However, it is advisable to take smaller depths of cut as the workpiece could slip on the mandrel.

A mandrel press

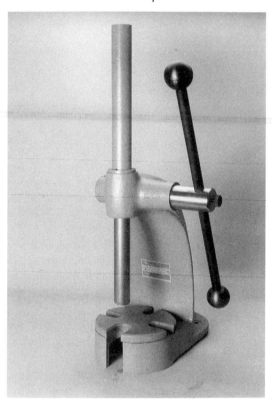

Courtesy of WDS Ltd.

A pulley mounted on a mandrel that is ready for turning between centres

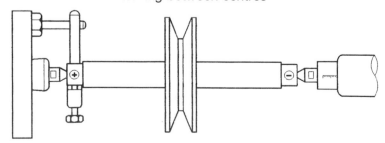

EXERCISE 9.5   Boring exercise

Make a plan to produce the boring exercise and indicate all tools required. Get your plan approved by your supervisor before starting work.

Your finished work should be within the drawing limits. On completion of the task, carefully measure your work and write in the table the actual sizes, noting any errors.

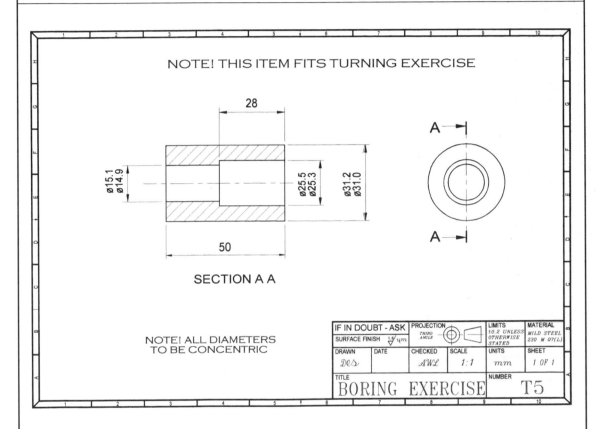

NOTE! THIS ITEM FITS TURNING EXERCISE

SECTION A A

NOTE! ALL DIAMETERS TO BE CONCENTRIC

IF IN DOUBT - ASK

PROJECTION	THIRD ANGLE		LIMITS ±0.2 UNLESS OTHERWISE STATED	MATERIAL MILD STEEL 230 M 07(L)
SURFACE FINISH				
DRAWN DES	DATE	CHECKED AWL	SCALE 1:1	UNITS mm	SHEET 1 OF 1
TITLE BORING EXERCISE				NUMBER T5

Planning:

Procedure	Tools and equipment selected
1. Acquire material and check for size.	*centre lathe*
2.	
3.	
4.	
5.	
6.	
7.	

Procedure (cont.)	Tools and equipment selected (cont.)
8.	
9.	
10.	

Research:

State the type of mount your lathe's spindle nose is fitted with.	
Describe how the catch plate was fitted.	
List all the work-holding devices available for the centre lathe you have been using and give an application for each.	
State where the machine's work-holding devices are stored so that they are not dangerous.	

Write here the safety **procedures** to be observed when changing the work holding device:

1.

2.

3.

Finished inspection report.

Complete the chart for all sections:

Component dimensions (mm)	*Limits:* (trainee to complete)	*Actual size:*	*Error:*
1. Overall length			
2. Outside diameter			
3. Bore diameter			
4. Bore depth			
5. Reamed hole diameter			

Results:

Is the workpiece to the drawing specification?   YES   NO

If completed workpiece is below drawing specification, list here the reasons for any errors, stating clearly how the errors will be avoided in future.

Start time and date:	End time and date:	Time taken (hours):

Witness testimony

I confirm that the boring exercise was made within the stated limits and that. . . . . carried out the work in a safe manner and changed the work-holding device correctly and safely.

Signed . . . . .Job title . . . . Date . . . .

Knurling

A knurl is a straight or a diamond-indented pattern that provides a 'thumb grip' on the outside of screwed components. A knurl is put on a surface with a knurling tool that has two small HSS wheels that are forced on the workpiece's surface, reshaping it to form the pattern.

Two types of knurling tools are commonly available as shown below:

Caliper-type knurling tool

Courtesy of WDS Limited

Pivot head or plunge-type knurling tool

Courtesy of WDS Limited

Method of using the caliper type:

(a) The knurling tool is secured in the lathe tool post with the jaws square to the axis of the work. If necessary, the work should be supported by the tailstock centre.

(b) Adjust the cross slide of the centre lathe to position the knurling wheels diametrically over the workpiece.

(c) With the cam handle in the vertical position, close the wheels by means of the adjusting screw until the wheels contact the workpiece. If a number of the same components are to be knurled, the adjusting screw should be locked with the locknut.

A caliper knurling tool operating on a precision centre lathe

Courtesy of WDS Limited

(d) With the centre lathe running at a moderate speed, pull the cam handle down until a distinct form appears on the workpiece. Traverse the wheels slowly along the workpiece using coolant.

(e) For higher-quality knurling, release the cam at the end of the first pass and repeat the operation – this time using light pressure and higher work speed. A clean well-formed knurl will result.

Tapping

Tapping (cutting internal threads with taps) can be carried out on a centre lathe:

1. by hand, using hand taps and a tap wrench or
2. using machine taps with special machine taps in a 'tapping head'.

1. Tapping on a centre lathe by hand

The process is as follows:

(a) Face-off the end of the workpiece.
(b) Centre-drill the end of the workpiece deep enough to chamfer the hole and provide a lead ready for tapping.
(c) Drill the workpiece with the appropriate tapping size drill (see Appendix III) to the required depth. If a 'blind hole' is being prepared, drill it 4 or 5 mm deeper than the minimum thread length.
(d) Switch off the electricity supply at the mains.

For taps with a centre hole:

(e) Insert a centre into the tailstock (if not see below).

(f) Select a suitable size tap wrench and mount the taper tap in the wrench.

(g) Apply tapping compound to the tap.

(h) Position the tool post so that its plain edge is facing the work-piece.

(i) Locate the tap wrench handle on the flat of the tool post and the end of the tap in the tailstock centre.

(j) Insert the front of the tap in the mouth of the drilled hole.

(k) Put the machine in neutral gear.

The set-up for hand tapping on a centre lathe

(l) Turn the chuck carefully by hand to cut the thread, relieving the chips frequently and maintaining **light** pressure on the tap with the tailstock.

(m) For deep holes, remove the tap occasionally to clear the chips.

(n) Manually turn the chuck backwards to remove the tap.

(o) Insert the second tap and finish the thread.

(p) Use the plug tap only to finish blind holes to full depth.

For taps without a centre hole, complete stages (a) to (d) as above and then continue as follows:

(e) Mount a drill chuck in the tailstock.

(f) Fit the taper tap in the drill chuck.

(g) Apply tapping compound to the tap.

(h) Insert the front of the tap in the mouth of the drilled hole.

(i) Put the machine in neutral gear.

(j) Turn chuck carefully by hand and allow the tailstock to slide along the bed (this starts the thread square to the workpiece's axis).

(k) After the first three or four threads are cut, remove the taper tap from the drill chuck.

(l) Transfer taper tap to a tap wrench and complete stages (m) to (p) for taps with a centre hole (above).

2. Machine tapping on a centre lathe

Special tap holders called tapping heads and machine taps are necessary for machine tapping.

Machine 'tapping head'

A machine 'tapping head' enables tapping to be carried out using a machine's power.

Special machine taps are needed when using a tapping head.

The tapping head holds the tap square to the workpiece that is being threaded.

Machine 'tapping head'

Courtesy of WDS Ltd.

Machine taps:

Machine taps are stronger than hand taps, having an angled cutting face that forces swarf to the front of the tap.

For cutting blind holes, a spiral helix tap like the one shown here should be used. Its flutes are designed to direct swarf out of the hole.

Machine tap

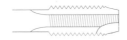

Courtesy of Kennametal
Hertel EDG Ltd.

*Spiral helix machine tap
(for blind holes)*

Courtesy of Kennametal
Hertel EDG Ltd.

The process for machine tapping is as follows:

(a) Face-off the end of the workpiece.
(b) Centre-drill the workpiece to the same diameter as the finished thread; this gives a slight lead.
(c) Drill the workpiece with the appropriate tapping size drill to the required depth; if a 'blind hole' is being prepared, drill it 4 or 5 mm deeper than the minimum thread length.
(d) Select the machine tap to suit the process.
(e) Consult tap manufacturer's data book and set centre lathe's spindle speed.
(f) Insert the tap holder into the tailstock's Morse taper barrel.
(g) Insert the tap in the tap holder.
(h) With the spindle operating, wind the tailstock forwards, entering the tap into the hole.
(i) The cutting process will draw the tap into the hole and when the cutting force becomes excessive or the tap 'bottoms', a friction clutch will allow the tap to turn in the 'tapping head'.
(j) Reverse the direction of rotation of the spindle and the tap will be extracted from the hole.
(k) If the tap becomes excessively tight, it can be removed, cleaned and more tapping compound applied. Repeat the process to finish tapping through the hole.

Parting off

When parts have been machined on the end of a long barstock, it is sometimes beneficial to cut the finished workpiece off the barstock in the lathe with a parting-off tool.

The parting-off tool is fed in behind the completed work and it cuts a groove separating the finished workpiece from the barstock.

The cutting tool used may be a special tungsten-tipped tool, a HSS-welded tip tool or an inserted blade type.

Parting off a workpiece in a centre lathe

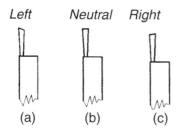

Left   Neutral   Right

(a)   (b)   (c)

The process is quite tricky because parting-off tools are relatively thin and fragile; consequently, special care is needed when parting off.

Parting-off tool tips are available 'handed'.

This handing refers to the slope on the tool's front edge. It either slopes to the left, so that the piece in the chuck is left with a pip on it or it slopes to the right, so that the piece that was cut off has a pip. Neutral tools are also available.

Before parting off

Always ensure the machine is as rigid as possible, this means:

- work is held securely in a chuck (you cannot part off work that is held between centres)
- the workpiece should protrude a minimum amount from the chuck
- the parting-off tool has a minimum overhang over the tool post
- the parting-off tool's blade length is only slightly more than the work's radius
- the compound slide is near the middle of its travel
- the saddle is locked to the bed with the screw on top of the carriage (where possible).

Always ensure that the machine is set up correctly – this means that:

- the parting-off tool is sharp and set at the correct centre height
- the spindle speed is slower than calculated for normal turning; half speed is advised
- the side of the parting-off tool is square to the workpiece; this prevents its sides 'binding' on the sides of the workpiece as it feeds into the groove
- Ensure that there is plenty of coolant available (except for cast iron, which does not need coolant).

Safety note! When parting off, it is particularly important to wear safety glasses and to concentrate well on what you are doing.

The procedure for parting off is as follows

(a) Line up the parting-off tool's right-hand edge with the required end of the workpiece.
(b) Start the spindle and feed the parting-off tool into the workpiece constantly and carefully.
(c) Reduce the feed rate as the parting-off tool point nears the centre of the workpiece.
(d) The workpiece will fall into the coolant tray.
(e) Turn off the spindle before recovering the workpiece from the coolant tray.

EXERCISE 9.6   Depth gauge parts

Plan the manufacture of the depth gauge parts listing the procedure and tools to be used. When your supervisor has approved your plan, make the three components in your planned sequence.

On completion of the task, carefully measure your work and write in the table the actual sizes, noting any errors.

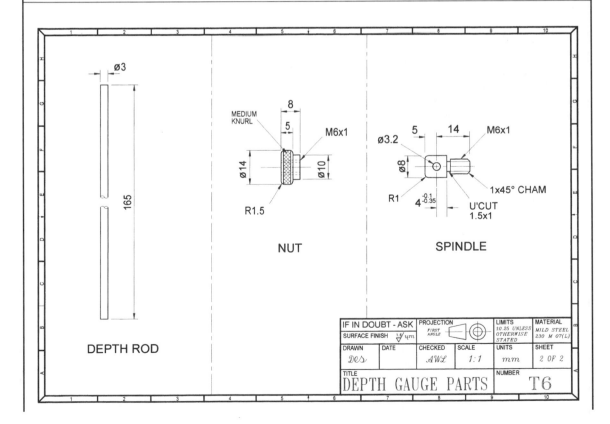

Planning:

Procedure
1. Acquire material and check for size.
2.
3.
4.
5.
6.
7.
8.
9.
10. Deburr and inspect for size.
11. Isolate and clean machine.

Tools and equipment selected

Research:

Write here the reason why components should not be removed from a three-jaw chuck until they have been checked and found to be of the correct size:	

List the range of turning tools that are available to you and state their typical applications:	Tool name	Application
	1.	
	2.	
	3.	
	4.	

	5.	
	6.	

Finished inspection report:

Complete the chart for all sections:

Component dimensions (mm)	*Limits:* (trainee to complete)	*Actual size:*	*Error:*
Spindle body length			
Spindle body diameter			
Nut length			
Nut ø10			
Thread assembly			
Rod length			

Results:

Is the workpiece to the drawing specification?

YES	NO

If completed workpiece is below drawing specification, list here the reasons for any errors, stating clearly how the errors will be avoided in future.

Start time and date:	End time and date:	Time taken (hours):

Witness testimony

I confirm that the turned depth gauge parts were planned out and made within the stated limits and that............................carried out the work in a safe manner and cleared the work area on completion of the task.

Signed................Job title................................Date..........

Setting work in a four-jaw chuck with a dial test indicator

When turning a long irregular workpiece in a centre lathe, the four-jaw chuck is the best work-holding device. The outside diameter can be turned and there is access to the end of the workpiece.

Care must be taken to set up the workpiece centrally in the chuck jaws so that the workpiece is located centrally. If you are making a workpiece that must be eccentric (e.g. a crankshaft), it can be held in a four-jaw chuck and set to the appropriate eccentricity.

Setting a workpiece in a four-jaw chuck

(a) Turn off the centre lathe at the mains electricity supply.
(b) Put the workpiece in the four-jaw chuck. Tighten the jaws so that they are all in a similar position relating to the concentric grooves on the chuck's front face. Stand a scribing block (or a surface gauge) on the machine's cross slide and adjust its point to touch the workpiece.
(c) Turn the chuck manually and observe which part of the workpiece is closest to the scribe, the high spot.
(d) Slacken the jaw nearest to the high spot and tighten the jaw opposite, moving the workpiece closer to the chuck's centre line.
(e) Repeat this process until the workpiece is touching the scribe all the time and all the chuck jaws are tight.

Surface gauge set on top of cross slide

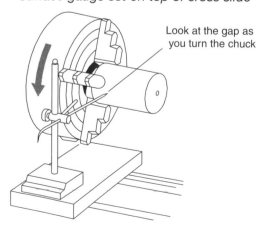

Look at the gap as you turn the chuck

Using a dial test indicator to set a workpiece accurately in a four-jaw chuck

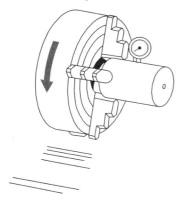

Note. To set the workpiece more accurately, a dial test indicator (DTI) can be used to measure any eccentricity.

(f) Remove the scribing block (or the surface gauge) and replace it with a DTI standing on a magnetic base.
(g) Set the DTI so that its stylus touches the workpiece.
(h) Slowly rotate the chuck by hand as before. You will see that the DTI's pointer shows the workpiece's eccentricity on its calibrated face.
(i) The work-setting procedure is the same as stages e and f above, but the sensitivity of the DTI will ensure that the workpiece is set more accurately. It is possible to set the workpiece concentric to within 0.02 mm.

Ensure that all the jaws are tight and continue as you would if you were working as normal.

EXERCISE 9.7   Toolmaker's clamp screws

Write a planning sequence for the manufacture of toolmaker's clamp screws and make two long screws.

On completion of the task, carefully measure your work and write in the table the actual sizes, noting any errors. Your finished work should be within the drawing limits.

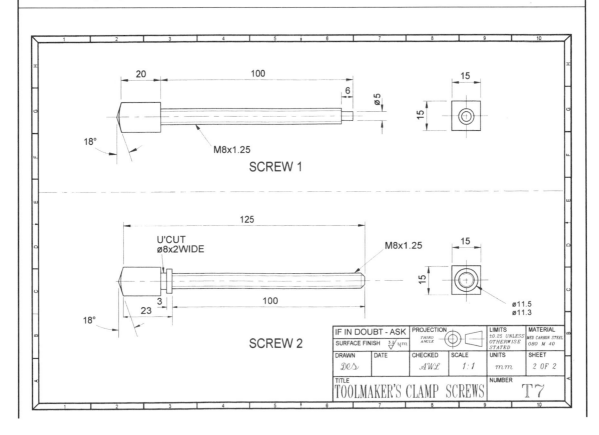

Planning:

Procedure	Tools and equipment selected
1. Acquire material and check for size.	
2.	
3.	
4.	
5.	
6.	
7.	
8.	
9.	
10. Isolate and clean machine.	

Research:

Find out what material the following tools are made from:

Tool	Material
Centre drill:	
Tailstock/headstock centre:	
Drills:	
Machine reamers:	
Screw cutting dies:	

Sketch:

Sketch here the set-up you used when setting the square bar run true in the four-jaw chuck using a DTI.

Finished inspection report.

Complete the chart for all sections:

Component dimensions (mm)	Limits: (trainee to complete)	Actual size:	Error:
Undercut width			
Undercut diameter			
Thread profile	*To fit thread gauge*		
ø5 mm end of Screw 1			

Results:

Is the workpiece to the drawing specification?   YES   NO

If completed workpiece is below drawing specification, list here the reasons for any errors, stating clearly how the errors will be avoided in future.

Start time and date:	End time and date:	Time taken (hours):

Witness testimony

I confirm that the toolmaker's clamp screws were made within the stated limits and that. . . . . carried out the work in a safe manner and cleared the work area on completion of the task.

Signed . . . . Job title . . . . Date . . . .

10 Machining engineering materials by milling

Exercise checklist

(Ask your assessor to initial the lower box for each exercise when you have completed it.)

Exercise no.	10.1	10.2	10.3	10.4	10.5
Initials					
Date					

Milling is the machining process generally selected for producing rectangular and angular workpieces with plane flat surfaces. Drilled and bored holes can also be accurately positioned with a milling machine. Milling involves feeding a securely held workpiece past a revolving multitooth cutting tool called a milling cutter. The sharp edges of the milling cutter remove material from the workpiece. The shape of the finished workpiece depends on:

- the movement of the workpiece relative to the cutter
- the shape of the cutter.

The operator of a milling machine is called a *miller*.

The column and knee design of milling machines is generally used for producing small quantities of components. Column and knee milling machines are available with:

- a vertical spindle, called a *vertical milling machine*
- a horizontal spindle, called a *horizontal milling machine*.

Vertical and horizontal milling machines have a similar lower construction, a solid 'column' and an adjustable 'knee'. Both types of machine are used to produce rectangular and angular workpieces with plane flat surfaces, and holes can also be drilled, bored and tapped. Vertical milling machines are more versatile and better suited to drilling while horizontal milling machines are more often used for heavy-duty work and production.

The principal differences between vertical and horizontal milling machines are:

Vertical milling machine   *Horizontal milling machine*

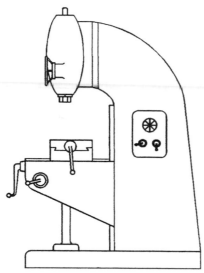
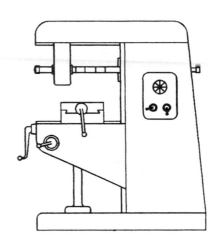

- the cutting tool is held in a different plane, depending on whether the machine has a vertical or a horizontal spindle
- different cutting tools are used
- the cutting tools are held in a different way.

For the exercises in this unit you will use both vertical and horizontal milling machines.

We start by examining the common types of vertical milling machines.

The vertical milling machine

Two different types of vertical milling machines are shown below. Both have a main frame called a *column* and a work table mounted on a knee that adjusts the table's height.

The workpiece is securely held to the machine's table in a work holding device, usually a vice or a special fixture. The cutting tool is fixed in the machine's spindle and rotated. The workpiece is traversed relative to the tool, thus removing material.

The fundamental parts of vertical milling machines are labelled on the following diagrams:

Vertical milling machine

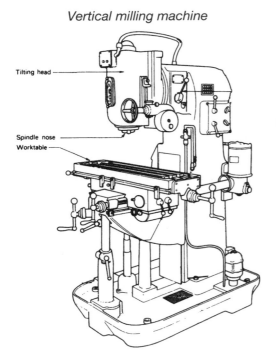

Turret milling machine

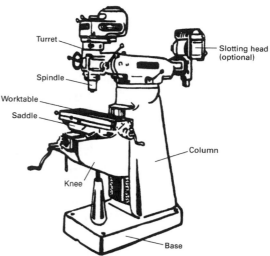

Courtesy of Bridgeport Machine Tools

- **The spindle** – The spindle is the rotating part of the machine, it houses the chuck and the cutting tool. The spindle's axis is vertical on all vertical milling machines. The spindle's bore is finely finished to a standard taper size so that standard tool-holding devices, that is, chucks, can be mounted. The spindle's speed can be set with levers and pulley belts to give a range of speeds.
- **The column** – The main frame of all milling machines is called the *column*. It must be rigid to withstand the cutting forces, so it is usually substantial in size and made in a box section from cast iron. The column on standard vertical milling machines contains the main motor and gearboxes to provide a range of speeds for table movement (feeds). At the foot of the column is a hollow base that provides a reservoir for the coolant and a platform for the machine to stand on.
- **The turret** (turret type only) – The head of the turret is mounted on a sliding beam and contains the spindle motor. The spindle can be raised or lowered with a lever in a similar way to a drilling machine. The turret head can be tilted both sideways and forward or backwards – this feature makes the machine very versatile, but less rigid than the standard type vertical milling machine.
- **The knee bracket** – The knee supports the machine's saddle so that the table is level and at 90° to the spindle. The knee is a large slide mounted on vertical dovetail slides at the front of the column. It is raised and lowered by a handle that operates the jacking screw.
- **The saddle** – The saddle slides across the top surface of the knee bracket and has another slideway across its top surface to enable the machine's table to be fed longitudinally.
- **The table** – The table is fed longitudinally or transversely and carries the workpiece past the revolving cutter. The table has a large accurately machined upper surface and tee slots used to mount the workpiece or work-holding device. The inside of the tee slots

Table movements

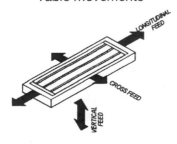

is machined to accurately locate work-holding devices via small 'pegs' called *tenons*.

- **Vertical milling machine sizes** – The size of a vertical milling machine is measured in two ways:

  - Table size
  - Traverse lengths.

The power of the motor will also be quoted in a machine specification.

Milling techniques used on vertical milling machines

Two basic techniques used to produce components on a vertical milling machine are shown below:

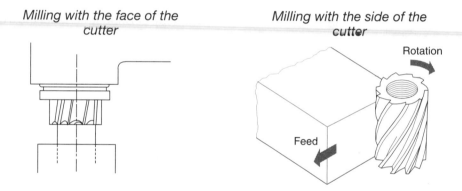

Milling with the face of the cutter

Milling with the side of the cutter

Work holding using a vice

A vice is a simple and efficient method of securely holding work-pieces of up to about 150 mm long on a milling machine's table. Machine vices may have two pegs fitted on the underside called *tenons*.

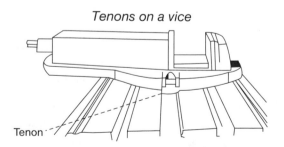

Tenons on a vice

Tenons locate the vice into one of the machine's tee slots, so the vice jaws are square to the machine's table. Vices with tenons can be set up quickly and accurately on the machine table. If tenons are not fitted to a vice, it must be set square accurately using a dial test indicator (DTI).

Plain machine vice

Courtesy WDS Limited

Swivel machine vice

Courtesy WDS Limited

Three types of machine vices are shown below.

- **The plain machine vice** – Used for holding workpieces square to the machine's table. Plain machine vices are strong and rigid, and when fitted with tenons they are quick to set up. A plain machine vice can be fitted to a swivel base that converts it into a swivelling machine vice.
- **The swivelling vice** – Used for holding workpieces to the machine's table when machining angular workpieces. The angle is set by swivelling the body of the vice. The angle of the vice can be read off a rotary scale that is calibrated in degrees. The vice is then locked that into position.

  For swivelling vices to be accurately used, they must be fitted with tenons that locate the vice on the machine's table so that when the vice is set at 0° the jaws hold the workpiece square to the table. If no tenons are fitted, the vice must be set up as described below.

- **Tilting and swivelling machine vice** – Used when machining compound angles, the workpiece can be turned around at an angle and tilted upwards as shown. This type of vice is used for complicated jobs because it needs to be set both horizontal and square planes before starting work.

  This type of vice is less rigid than those above and can take time to set up accurately with a DTI.

Tilting and swivelling machine vice

Courtesy WDS Limited

Pull down vice jaws

Courtesy WDS Limited

- **Pull down vice jaws** – These jaws can be fitted to any of the above machine vices. They effectively pull the workpiece down to the vice base or packing, eliminating the need to tap the workpiece down after tightening the vice. Pull down vice jaws may have plain or serrated faces.

Safety note! Always use correct lifting technique when lifting heavy machine vices. This means bending at the knees and keeping your head up so that your spine is vertical.

Always ask for help or use powered lifting equipment with loads that are too heavy or bulky for you to lift by yourself.

Correct lifting posture

Setting a vice square in a milling machine's table

When the milling square faces a workpiece, it is essential that the vice is mounted square on the machine's table. A machine vice fitted with tenons will always locate in this way but a vice without tenons should **always** be set square before any metal cutting can start. Setting of the vice square can be carried out with a DTI mounted on a magnetic base.

The procedure for setting a vice square is as follows

(a) Thoroughly clean the top of the machine table and the underside of the vice.
(b) Lightly secure the machine vice to the machine table with two forged tee bolts, nuts and washers.
(c) If using a swivel vice, carefully set the vice at 0°.
(d) Hold a parallel bar in the vice jaws as shown.
(e) Mount a DTI on a magnetic base.
(f) Attach the magnetic base securely to the machine's column so that the DTI's anvil rests against one end of the parallel bar.
(g) Wind the table lengthways so that the DTI's pointer shows any reading out of parallel. Note this reading.
(h) Tap the vice body to compensate for any 'out of parallel' DTI reading.

(i) Tighten the vice's securing nuts.
(j) Finally, check the vice in its fixed position for parallel. The DTI should have less than 0.05 mm total indicator reading (TIR).

Mounting workpieces in a vice

(a) When the machine vice is set square to the machine table the workpiece can be clamped to the machine vice. The predominant direction of the cutting force should always be towards the **fixed jaw** of the vice (as opposed to the moving jaw).

Workpiece mounted on parallel bars in a vice

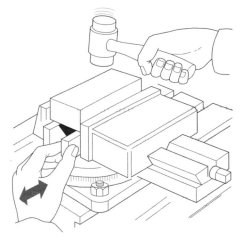

(b) The workpiece is then mounted on a pair of parallel bars and the vice is secured with the handle.
(c) Tap the workpiece down with a hide hammer until both parallels are trapped tightly between the top of the vice and the workpiece. You now know that the underside of the workpiece is parallel to the top of the vice.

After being secured, the workpiece is slowly fed under a rotating milling cutter. The cutter should be carefully selected for the particular job in hand.

Selection of milling cutters for vertical milling

When selecting a milling cutter for vertical milling operations, choose a milling cutter that has the largest possible diameter. This is to maintain the cutter's rigidity.

The most common types of vertical milling cutter are described below:

- **End mill**: Made from high speed steel (HSS), these cutters are designed for profile milling. They can also be used for producing open slots. Close tolerance slots can only be produced with a cutter which is smaller than the slot. The slot is then accurately finished by making minor adjustments.

End mill

- **Slot drill**: Made from HSS, these cutters can be used to produce slots and keyways accurately. They are centre cutting and can plunge feed vertically into a workpiece to produce closed slots or pockets as one tooth extends to the centre of the tool.

Slot drill

Other specially shaped HSS cutters for vertical milling:

Ball nose cutting tool

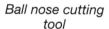
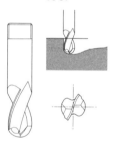

Designed to produce 3D profiles

Tee slot cutting tool

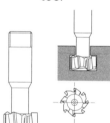

For opening out slots into tee slot shape

Dovetail cutter

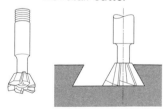

Produces a slot for a woodruff key in shafts

Ripping cutter

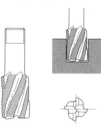

For rapid rates of material removal

Courtesy of Kennametal Hertel EDG Ltd.

Large diameter cutters for vertical milling:

• **Shell end mill:** Large cutter used for facing large flat surfaces. Shell end mills are mounted on a tool holding device called a stub arbor. The HSS cutting 'shell' is fixed onto the stub arbor with a screw. Positive drive is ensured by two slots in the shell locating with driving lugs on the stub arbor.	*Shell end mill* 	• **Face mills:** Face mills fit directly into stub arbors fitted in the machine's spindle, which giving maximum rigidity. They are fitted with indexable HSS or tungsten carbide inserts that make them very hard. Although this type of cutter is very expensive, the hard and rigid combination enables faster cutting speeds to be used for more rapid production rates.	*Face mill* 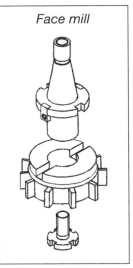

Mounting a chuck on a vertical milling machine

Chucks are the most common cutting tool holding device used for vertical milling operations. Before you mount the cutting tool, you must fit a chuck in the spindle correctly.

The procedure for mounting a chuck on a vertical milling machine is as follows:

(a) Clean the inside of the machine's spindle and the tapered end of the chuck with a clean cloth.

Inserting a chuck in a vertical milling machine

(b) Reach up to the top of the machine's spindle and hold the drawbar.
(c) Insert the chuck's taper into the spindle's tapered hole.
(d) Locate the chuck in the driving lugs on the spindle nose.
(e) Screw the drawbar into the end of the chuck and tighten with the locknut.

Mounting the cutter on a vertical milling machine

Milling cutters are sharp and should be carefully handled with gloved hands. Always isolate the machine before changing the milling cutter; should the machine be accidentally started, a serious accident could result.

The usual chuck used on a vertical milling machine for holding end mills and slot drills is similar to the Dormer Fastloc chuck shown below. This type of chuck is designed to hold screwed shank cutters efficiently.

Component parts

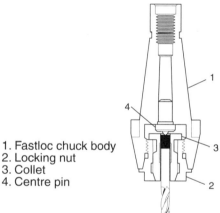

1. Fastloc chuck body
2. Locking nut
3. Collet
4. Centre pin

Courtesy of Dormer Tools Limited

The chuck incorporates all the advantages of the self-locking technique for screwed shank cutters: a hardened and ground centre and a screwed precision collet that is free to rotate with the locking nut for automatic tightening under heavy cuts.

The internal locking nut has a precision ground taper location with the chuck body to maintain concentricity even after years of use.

Assembly instructions for mounting end mills or slot drills on a milling chuck

(a) All component parts must be cleaned before assembly, particularly at the ground taper locations.
(b) The collet should be inserted into the locking nut, making sure that the driving flats are engaged in the locking nut flats.
(c) The collet and locking nut assembly should then be inserted into the chuck body and screwed in until the mating tapers of the locking nut and the chuck body are engaged.

(d) The selected cutter should then be inserted into the collet and screwed in by hand pressure until it firmly locates on the centre pin. Wear protective gloves for this operation.

(e) Tighten the locking nut with a special spanner, ensuring that the mating tapers are securely located.

To release the cutter: Using the ring spanner provided, release the locking nut slightly until the cutter can be unscrewed from the collet by hand.

Selection of spindle speeds for milling

The speed at which the machine spindles, and hence the cutter, depends on four factors:

- the diameter of the cutter
- the type of material being cut
- the material from which the cutting tool is made
- the availability of coolant.

To calculate the correct speed with the correct cutting fluid, use the formula:

$$\text{rev/min} = \frac{1000 \times S}{\pi \times d}$$

Where S = cutting speed of the work material
                 d = cutting tool diameter
                 π = 3.142 (often approximated to 3)

See Appendix VIII for the cutting speeds for various tool/workpiece combinations to use as a guide. Some cutting tool manufacturers recommend speeds different from those quoted in Appendix VIII.

Selecting feed rate

The feed rate for milling cutters should be carefully selected. Feed rates for milling are usually stated as mm per minute. The feed rate depends on the spindle speed, the number of teeth on the cutter and the feed per cutter tooth. The correct feed rate for a milling operation can be calculated by referring to Appendix IX to find the recommended feed rate per tooth and using the formula:

Feed per minute = Rev/min × No. of teeth on cutter × Recommended feed/tooth

Direction of feed

The two views shown below are alternative methods of removing material along the side of a milling cutter. Although both methods are practised in theory, up cut milling should normally be used.

Up cut milling
Recommended for most conventional operations

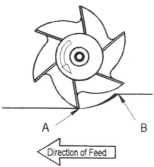

A   B

Direction of Feed

Down cut milling
Only for special machines

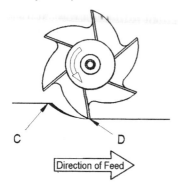

C   D

Direction of Feed

Study the up cut Milling drawing carefully. Note that the direction of the cutting tool's point is opposite to the direction in which the workpiece is being fed. The chip is cut at 'Point A' and is parted from the workpiece at 'Point B', that is, the chip formation starts thin and ends thick.

When using the up cut milling technique, the opposing directions of movement of tool and work prevent the cutter from snatching the work in if there is any slack between the leadscrew and its driving nut (backlash).

The down cut milling method (sometimes called climb milling) will give a better surface finish because the chip starts thick at 'Point C' and finishes thin at 'Point D'. Down cut milling can be used for heavier cuts because each tooth passing the workpiece pushes the workpiece down to the parallels.

Although down cut milling offers these advantages, it is only suited to special machines with backlash eliminators and should not be used on general tool room machines.

Depth of cut

When removing material by milling, the depth of cut selected is dependent on the same factors as that for turning. These include:

- type of cutting tool material
- cutting tool condition and geometry
- type of material being cut
- availability and type of coolant
- feed rate set
- power of the machine's motor
- general condition of the machine
- rigidity of the machine's tool post and work-holding device.

If there is any vibration, stop the machine immediately, check if the tool is sharp and reduce the depth of cut and/or the feed rate.

Positioning of guards

Milling is an extremely dangerous process. Never touch a rotating cutter or attempt to clear away swarf while the machine is running. It is essential that all machine guards are positioned and used appropriately to offer maximum protection to the operator.

For vertical milling machines, the enclosure mill guard is typical. The guard is transparent; and in the closed position, protects the operator from flying debris and being splashed by coolant. More importantly, the guard prevents the operator from becoming entangled with any moving parts.

When opened, this type of guard offers unrestricted access to the machine's table and work area. Guards of this type are very efficient as they are easy to fit, guard the whole of the cutting area and are transparent.

Enclosure mill guard closed and ready to operate

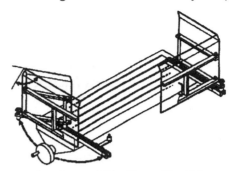

Courtesy of Silvaflame Co. Ltd.

Electrical interlocks are sometimes fitted with this type of guard to prevent the machine from being started until the guard has been positioned and locked into place.

The Health and Safety at Work Act 1974 states that you must take reasonable care of your health and safety and that of others who may be affected by your acts. This means you must **always** use guards and report problems such as failure to operate properly, damage to guards or missing guards to your supervisor.

Operating the vertical milling machine

Never work on any machine until you have been shown how to operate it properly by a suitably qualified person.

Your supervisor will arrange for you to be shown how to operate a vertical milling machine and introduce you to the machine's features, including:

- power supply and how to isolate the machine when changing and cleaning cutting tools
- guards and guarding systems to be used when operating the machine
- spindle stop, start and reverse controls
- hand wheels and feed dials for longitudinal, vertical and cross feed
- spindle speed selection levers
- feed selection levers
- feed engagement levers and trips for 'over travel'
- power feed control for traversing the work
- rapid feed (if available)
- locks on all three slides
- coolant/cutting fluid supply
- how to manage 'backlash'
- emergency stop procedure.

If you do not understand what you have been shown or if you are unsure how to proceed you must *ask*!

Milling is dangerous!

Always use guards and do not take risks.

Never operate a machine until the slideways are cleared of all tools and equipment.

When you have practised working the vertical milling machine using pieces of spare material and are confident of the controls, ask your supervisor for permission to start Exercise 10.1, the stepped block.

EXERCISE 10.1   Stepped block

Task (this task was marked out in Exercise 7.4)
Make the stepped block using a vertical milling machine. Use the procedure and tools indicated. Your finished workpiece should be within the drawing limits. On completion of the task, carefully measure your work and write in the table the actual sizes, noting any errors.

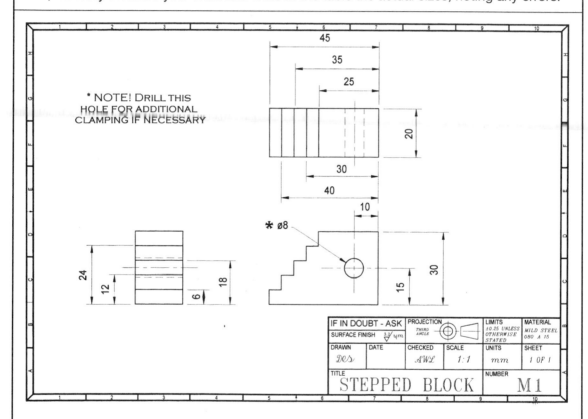

* NOTE! DRILL THIS HOLE FOR ADDITIONAL CLAMPING IF NECESSARY

* ø8

IF IN DOUBT - ASK	PROJECTION THIRD ANGLE		LIMITS ±0.25 UNLESS OTHERWISE STATED	MATERIAL MILD STEEL 080 A 15
SURFACE FINISH 3.3 μm				
DRAWN DES	DATE	CHECKED AWL	SCALE 1:1	UNITS mm	SHEET 1 OF 1
TITLE STEPPED BLOCK			NUMBER M1	

Planning:

Procedure	*Tools and equipment selected*
1. Acquire material, check for size, and deburr all sharp edges. 2. Hold work in a plain machine vice on parallel bars. 3. Mount end mill in chuck; set spindle speed and feed rate. 4. Machine ends square and to length. 5. Mark-out guidelines on workpiece blank. 6. Set up workpiece vertically in vice.	*Vertical milling machine* *End mill* *Collet chuck* *Machine vice and parallel bars* *Hide hammer* *Metric 300 mm rule* *0–25 mm micrometer* *25–50 mm micrometer*

Procedure (cont.)	Tools and equipment selected (cont.)
7. Touch on workpiece top and side, setting registers at zero. 8. Machine all steps on workpiece. 9. Deburr all sharp edges. 10. Stamp your name on your work. 11. Isolate machine from the electricity supply. 12. Clean the work area and machine.	*Letter stamps*

Examine the **vertical milling machine** you used and complete the table with the required information:

Make of machine:	
Model of machine:	
Type of vertical milling machine (standard or turret):	
Number of spindle speeds available on the machine:	
Selected spindle speed for mild steel using a ø25 HSS end mill (show your calculation):	
Selected feed rate mild steel using a ø25 mm 4 tooth HSS end mill (show your calculation):	

Write here the emergency stop procedure for the vertical milling machine:

1.

2.

3.

Write here why it is important to isolate the machine before changing cutting tools and before cleaning the machine:

Describe, using a sketch if necessary, how the work holding device was

1. mounted on the machine
2. secured to the machine
3. aligned correctly.

Finished inspection report:

Component dimensions (mm)	Limits:	Actual size:	Error:
Height 45	45.25 44.75		
Height 35	35.25 34.75		
Height 25	25.25 24.75		
Width 18	18.25 17.75		
Width 24	24.25 23.75		

Results:

Is the workpiece to the drawing specification?   YES   NO

If completed workpiece is below drawing specification, write here the reasons for any errors, stating clearly how the errors will be avoided in future.

Start time and date:	End time and date:	Time taken (hours):

Witness testimony

I confirm that the stepped block was made within the stated limits and that
. . . . carried out the work in a safe manner, using appropriate
roughing and finishing cuts, clearing the work area on completion of the task.

Signed . . . . Job title . . . . Date . . . .

Cutting angular faces

When angles are specified on a drawing, the workpiece can be set up in a vice at the required angle and then fed under the cutter. A quick and reliable method of setting up angular workpieces for machining is in a:

- rigid machine vice, with the workpiece **inclined** at the required angle or;
- swivel machine vice, with the workpiece **turned** to the required angle.

Procedure for setting workpiece in a rigid machine vice inclined at an angle

Isolate the machine's electrical supply.

*Setting the workpiece at 27°
in a rigid machine vice*

(a) Select a rigid machine vice with tenons that fit the table's tee slots.
(b) Thoroughly clean the top surface of the milling machine table and the bottom surface of the machine vice.
(c) Clamp the vice securely to the machine table.
(d) Check if the vice is fitted square.
(e) Mark-out the workpiece.
(f) Clamp the workpiece between the vice's jaws.
(g) Adjust a combination set's protractor head at the required angle.
(h) Rest the protractor against the workpiece and tilt the workpiece until the spirit level indicates level.
(i) Tighten the workpiece securely in the vice.
(j) Mount all machine guards.

Procedure for setting a workpiece in a swivelling vice turned at the required angle

(a) Isolate the machine's electric supply.
(b) Select a swivel type machine vice with tenons that fit the table's tee slots.

Mounting a workpiece at 27° in a swivel vice

Set at 63°

(c) Thoroughly clean the top surface of the milling machine table and the bottom surface of the machine vice.
(d) Clamp the vice securely to the machine table.
(e) Slacken the locking nuts on the swivel base of the vice and turn the vice through the required angle.
(f) Lock the vice at the required angular setting.
(g) Check if the angle is correct.
(h) Tap the workpiece down tight on the parallel bars, with a hide hammer.
(i) Make sure the parallel bars will not interfere with the cutting process.
(j) Make sure there is room for the cutter to clear the workpiece after it has cut the material.
(k) Mount all machine guards.

Machining angular workpieces

(a) Mark-out the finished angle on the workpiece and mount the workpiece in a machine vice as outlined above.
(b) Fit the chuck in the machine spindle and mount an end mill.
(c) Select and set the correct spindle speed and feed rate.

Milling the angular workpieces in a rigid vice with a face mill

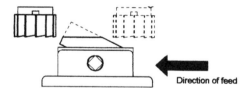

Direction of feed

(d) Position the machine table so that the side of the cutter is ready to cut the material using up cut milling.
(e) Mount **all** guards.
(f) Turn on the electric supply.

Milling angular workpieces in a swivel vice with an end mill

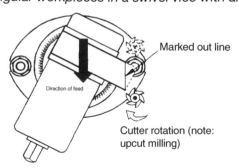

Marked out line

Direction of feed

Cutter rotation (note: upcut milling)

(g) Machine the surface using appropriate coolant.
(h) On completion of the machining process, isolate the machine and carefully remove the workpiece.

Slot drilling

Slot drilling is the process of cutting closed slots into workpieces. To make such shapes a slot drill must be used because it has a cutting edge extending to the tool's centre, allowing it to be fed vertically down into the workpiece, i.e. plunged.

(a) Mark-out guidelines on the workpiece.
(b) Mount the work-holding device square on the milling machine's table.
(c) Fit a slot drill that has a diameter smaller than the width of the required slot. The smaller cutter allows the slot to be accurately 'opened out' to width.
(d) Set the appropriate machine's spindle speed and feed rate.
(e) Position the table so that the milling cutter is nearly touching the side of the workpiece.
(f) Set the table's hand wheel reading to zero.
(g) Move the table slowly and carefully until the cutter **just** touches the side of the workpiece (some millers do this with the machine isolated using a 0.05 mm feeler gauge).
(h) Lower the table so that the end of the milling cutter is clear of the workpiece.
(i) Move the table sideways one half of the cutter's diameter so that the centre of the cutter is in line with the edge of the workpiece.
(j) Reset the table's hand wheel reading to zero.
(k) Now you can move the table sideways, to an appropriate distance on the drawing, so that the cutter's axis is exactly above the slot centre line.
(l) Slowly plunge the cutter downwards into the workpiece to a maximum depth equal to the cutter's diameter.
(m) Feed the workpiece sideways to cut the slot to the correct length.
(n) Repeat stages l and m by applying cuts by plunging the cutter slowly into the workpiece and feeding along the length of the slot until the full depth is reached.
(o) Measure the width of the sides of the workpiece. Wind the table sideways to open out the slot if necessary, to correct any positional error. Remember to 'up cut mill' when doing this.

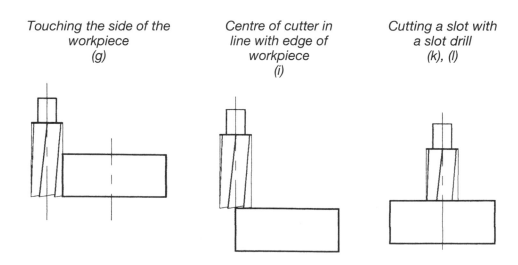

Touching the side of the workpiece
(g)

Centre of cutter in line with edge of workpiece
(i)

Cutting a slot with a slot drill
(k), (l)

EXERCISE 10.2   Slotted Clamp

Make the slotted clamp using a vertical milling machine. Use the procedure and tools indicated. Your finished workpiece should be within the drawing limits. On completion of the task, carefully measure your work and write in the table the actual sizes, noting any errors.

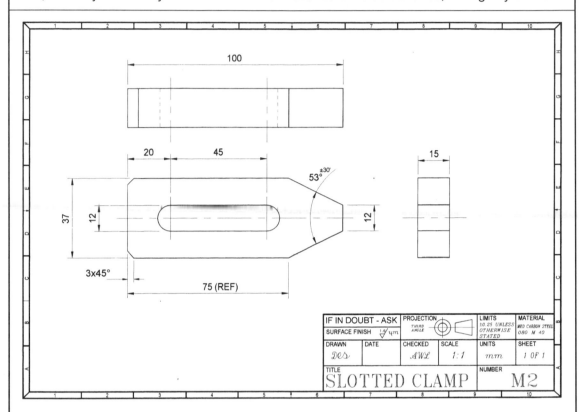

IF IN DOUBT - ASK	PROJECTION		LIMITS	MATERIAL
	THIRD ANGLE		±0.25 *UNLESS OTHERWISE STATED*	*MED CARBON STEEL* 080 M 40
SURFACE FINISH 1.6/μm				
DRAWN	DATE	CHECKED	SCALE	UNITS	SHEET
Des		*AWL*	1:1	*mm*	1 OF 1
TITLE				NUMBER
SLOTTED CLAMP				M2

Planning:

Procedure	Tools and equipment selected
1. Acquire material, check for size. 2. Mount the vice on the machine and set square. 3. Set the workpiece on the vice using parallel bars. 4. Mount the end mill and set speed and feed. 5. Machine blank to size. 6. Mark-out guidelines on workpiece. 7. Mount workpiece at 27° and machine each angle, reset workpiece at 45° and machine the corner chamfers. 8. Change the cutter for the slot drill. 9. Machine the slot in position. 10. Isolate machine from the electric supply.	*Vertical milling machine* *End mill* *Slot drill* *Collet chuck* *Machine vice and parallel bars* *Hide hammer* *300-mm rule* *0–25-mm micrometer* *25–50-mm micrometer* *Vernier caliper* *Letter stamps*

Procedure (cont.)	*Tools and equipment selected (cont.)*
11. Deburr all sharp edges and stamp your name on work. 12. Clean the work area and the machine.	

Complete the tables with correct information:

Depth of cut used for roughing out cuts with: 1. end mill	
2. Slot drill	
Depth of cut used for finish cuts with: 1. end mill	
2. slot drill	
Name the type and supplier of coolant used in your milling machine and, if soluble oil is used, state the diluting ratio.	
How can distortion due to stress in the material be avoided?	

Sketch Make a sketch (or series of sketches) to show how the cutting tool was set over the workpiece's datum.	

Explain in your own words why it is necessary to check the workpiece thoroughly before removing it out of the work holding device.

Finished inspection report:

Component dimensions (mm)	Limits:	Actual size:	Error:
Body thickness 15	15.25 14.75		
Slot width 12	12.25 11.75		
Overall max slot length 57	57.25 56.75		
Overall clamp width 37	37.25 36.75		
Angle 53°	52° 30′ 53° 30′		

Results:

Is the workpiece to the drawing specification?   YES   NO

If completed workpiece is below drawing specification, write here the reasons for any errors, stating clearly how the errors will be avoided in future.

Start time and date:	End time and date:	Time taken (hours):

Witness testimony

I confirm that the slotted clamp was made within the stated limits and that
. . . . carried out the work in a safe manner, managing backlash in the machine slides.
The trainee is aware of the range of operations the vertical milling machine is capable of performing.

Signed . . . . Job title . . . . Date . . . .

There are two types of horizontal milling machine:

- Plain horizontal milling machine (below left).
- Universal milling machine that has a swivelling table (below right).

Horizontal milling machines

The two principal differences between vertical and horizontal milling machines are the orientation of the machine's spindle and the type of cutting tools used.

Knee and column horizontal milling machines

Plain horizontal milling machine

Universal milling machine

Courtesy of Cincinnati Milacron

Horizontal milling machines are very rigid and robust. They are used to produce rectangular and angular workpieces with plane flat surfaces. The plain horizontal milling machine is a less versatile machine than the vertical milling machine. On a horizontal milling machine, a multitooth cutting tool is fixed onto the standard tool-holding device called the arbor. The arbor is a long shaft fitted into the spindle of the machine by a taper, it is supported by a yoke.

As with vertical milling, the workpiece is securely held in a work-holding device that is usually a machine vice or a special fixture. The workpiece is fed relative to the rotating cutting tool, thus removing material. The swivelling table on the universal milling machine enables the machine to be used for machining complex shapes and angles using various attachments. The horizontal milling machine is to be used for the next two exercises in this workbook.

The fundamental parts of a horizontal milling machine are labelled in the above diagrams and described below:

- **The spindle** – The rotating part of the machine into which the arbor and cutting tool is mounted. Its bore is finely finished to a standard taper size so that the arbor can be mounted. The spindle's speed can be set with levers and pulley belts. Note that the spindle is in a horizontal plane on the horizontal milling machine.

- **The column** – The frame of the machine must be rigid to withstand the cutting forces. The column houses the main motor and gear-boxes and provides a range of speeds for table movement (feeds). The machines shown above are rigid production machines. At the foot of the column is a base that provides a reservoir for the coolant and a platform for the machine to stand on.

- **Overarm** – The overarm is a rail on the top of the machine. It has a guideway that houses one or sometimes two yokes. When setting up a machine, you should always have the overarm retracted back as far as possible to reduce deflection of the arbor.

- **Yoke** – The yoke is mounted on the overarm and supports the arbor, reducing the chances of bending the arbor and providing rigidity and stability to the cutting tool.

- **The knee bracket** – The knee supports the machine's saddle so that the table is level and at $90°$ to the spindle. The knee is a large slide mounted on vertical dovetail slides at the front of the column – it is raised and lowered by a handle that operates the jacking screw.

- **The saddle** – The saddle slides transversely along the top surface of the knee bracket. It has slideways across its top surface to enable the machine's table to be fed longitudinally. The saddle on the universal milling machine is split in the middle to allow the table to be swivelled at an angle.

- **The table** – The table has a large accurately machined horizontal upper surface that is used to mount the work-holding device or workpiece. A series of accurately machined tee slots are used for clamping. The inside of these tee slots is machined to accurately locate work-holding devices via 'tenons' fitted to the underside of vices etc. The table is fed longitudinally or transversely, and carries the workpiece past the revolving cutter.

Horizontal milling machine sizes

The size of a milling machine is measured in two ways:

- Table size.
- Traverse lengths (i.e. the maximum movement the table can be traversed).

The power of the motor is also quoted in the machine's specification.

Basic techniques used to produce components on a horizontal milling machine are shown below:

Basic horizontal milling operations

Milling with face of cutter

Milling with side of cutter

Milling with cutter mounted in chuck or stub arbor

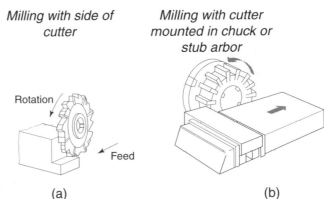

(a)   (b)

Note. Vertical milling machines are preferred for drilling holes because the operator has a better view of the workpiece.

Work holding in a horizontal milling machine

When setting up a workpiece in a horizontal milling machine, the workpiece should be mounted on a work-holding device in the same way as it would be for a vertical milling machine. For information on setting and selecting vices, see pages 316 to 319.

Selection of cutter for horizontal milling

There is a greater choice of cutters for horizontal milling than vertical milling. The cutter rotates and removes material as the workpiece traverses past it. There are three basic types of cutters, most are made out of HSS, although carbide tipped tooling is used on some types of tools.

- Plain milling cutters.
- Form relieved cutters.
- 'Chuck' and 'stub arbor' mounted cutters.

Plain milling cutters

- **Slab mill** (cylindrical cutters) – Used for cutting on the outside edge only (the periphery). Slab mills are used for machining wide flat surfaces. Helical (spiral) teeth have the advantage of more than one tooth cutting at a time – this reduces chatter and vibration.
- **Side and face cutter** – Has teeth on both sides and on its periphery. It is used to machine the face of workpieces and to take vertical cuts. It is normally made from HSS, although side and face cutters are also available with inserted carbide tips. It is not used for cutting slots in one pass as vibrations cause the sides of the cutter to cut the slot over size.

Slab mill

Side and face cutter

Slotting cutter

Slitting saw

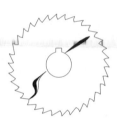

- **Slotting cutter** – Available in set sizes, slotting cutters are used to cut slots accurately to a specified width in one pass. They have teeth on their periphery but not on their sides. The width of a slot cut by a slotting cutter is the same as the width of the cutter.
- **Slitting saw** – This type of cutter is used to cut material to length and to cut narrow slots, of width between 1 mm and 5 mm, through a workpiece. Slitting saws are large diameter cutters (up to 200 mm), and have only teeth on their periphery.

Form relieved cutters

(a) **Single angle cutter** – Has specified angle on one face. The angle on this type of cutter enables the machining of angular faces and chamfers on corners. When using angle cutters, avoid heavy loads on the points of the cutting edges as they are easily chipped.

(b) **Double angle cutter** – Has specific angles on both faces. Often used for machining items such as splines and milling cutters.

(c) **Convex cutter** – Used to machine rounded slots and grooves in the surface of workpieces. 'Convex' refers to the shape of the cutter.

(d) **Radius (corner rounding) cutter** – For finishing off corners on large workpieces. A different cutter is required for each radius, so this is only used to finish off components when large quantities are required. Radius cutters are also available, with a full radius for rounding the ends of workpieces.

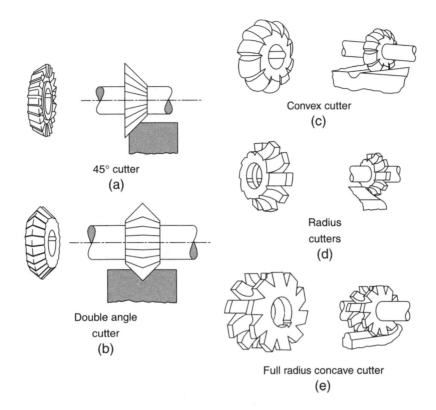

45° cutter
(a)

Double angle
cutter
(b)

Convex cutter
(c)

Radius
cutters
(d)

Full radius concave cutter
(e)

Chuck and stub arbor mounted cutters

These cutters are the same as those used in vertical milling. They are mounted in exactly the same way (see page 321).

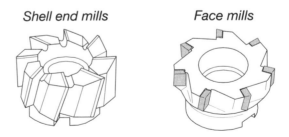

Shell end mills

Face mills

Fitting arbor mounted cutters in a horizontal milling machine

Plane and form relieved milling cutters are mounted on the machine's arbor. The position of the cutter is adjusted by selecting and positioning spacing collars on either side of the cutter. It is advisable to mount the cutter as close as reasonably possible to the machine's column to maintain rigidity. The yoke is used to support the arbor and runs on a larger collar that acts as a bearing.

Milling arbors are tool-holding devices and are manufactured from high tensile alloy steel for strength. They have an accurately finished parallel cylindrical surface and a tapered end that fits into the machine's spindle. Milling arbors are located in the spindle nose's driving pegs and taper, and then secured in position with a draw bolt that screws into the end of the arbor.

Spacing collars are for positioning the cutter and are made from case-hardened steel. The length of spacing collars is standardised and the ends are accurately ground parallel. One collar, called a **bearing bush**, has a larger diameter than the others. The bearing bush fits inside the yoke and steadies the arbor during cutting to prevent it from bending.

A large nut is fitted to the end of the arbor for holding the collars, the milling cutter and the bearing bush in position. The nut may have a left or a right hand thread.

Procedure for mounting arbors and cutters on a horizontal milling machine is as follows

(a) Clean the inside of the machine's spindle and the tapered end of the arbor with a clean cloth.
(b) Remove the yoke from the overarm.
(c) At the back of the machine, insert the drawbar into the spindle.
(d) Insert the arbor's taper into the tapered hole of the spindle nose.
(e) Screw the drawbar into the end of the arbor and tighten with the locknut.

Mounting the arbor

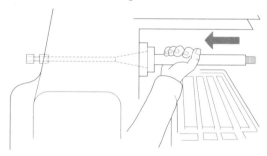

(f) Clean the arbor's outer surface and the inside and sides of all spacing collars.
(g) Put spacing collars on the arbor to locate the cutter.
(h) Fit a short key into the arbor's keyway.

  Note: (i) The key must **not** extend beyond the sides of the cutter.
      (ii) A key is not required for cutters under 6-mm thick.

(i) Select the direction of rotation of the cutter.

  • The cutter should rotate in the direction **opposite** to the arbor's nut when the nut is being tightened.
  • The cutter's rotation should be **against** the vice's fixed jaw or solid abutment of work holding device.

(j) Mount the cutter close to the column end of the machine to maximise rigidity, **use gloves or cloth to protect your hands** when handling milling cutters.

Mounting the cutter

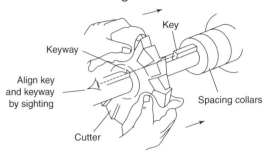

Note the short key

(k) Fit more spacing collars to position the cutter in a rigid position.

(l) Fit the bearing bush as close to milling cutter as possible to lessen the chances of bending the arbor.

(m) Fit more spacing collars to 'fill' the arbor up to the thread on the end.

(n) Fit and tighten yoke in position.

(o) Tighten the nut on the end of the arbor.

(p) Mount the cutter guard.

Slab milling cutter mounted correctly

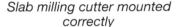

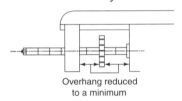

Overhang reduced to a minimum

Note. Cutters should not be mounted as shown below because:

(i) the arbor may bend

(ii) the cutter is not held rigid, causing chatter and resulting in a poor surface finish and inaccurate work.

Slotting cutter mounted badly

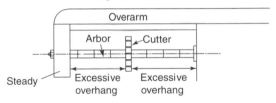

Positioning guards

Horizontal milling is an extremely dangerous process. You must never touch a rotating cutter or attempt to clear away swarf while the machine is running. It is essential that all the machine guards are positioned and used appropriately to offer maximum protection to the operator.

The guard used for vertical milling is also suitable for horizontal milling machines, but cutter guards are also frequently used.

A popular type of cutter guard for horizontal milling machines is shown. The guard is of steel construction and can be quickly adjusted to fit around the milling cutter. It will protect the operator from flying debris and prevent him from becoming entangled with moving parts.

You must **always** use guards provided on machine tools. Report any problems such as failure to operate properly, damage to guards, and so forth, to your supervisor. The requirements of the Health and Safety at Work Act 1974 require you to do so.

Cutter guard fitted to a horizontal milling machine

Courtesy Silvaflame

EXERCISE 10.3   Horizontal milling fixture

Task:

Make the milling fixture using a horizontal milling machine, following the procedure and tools indicated. Your finished work should be within the drawing limits. On completion of the task, carefully measure your work and write in the table the actual sizes, noting any errors.

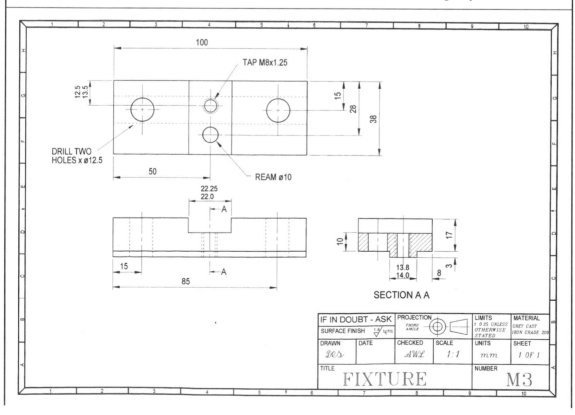

Planning:

Procedure	Tools and equipment selected
1. Acquire material and check for size. 2. Hold workpiece horizontally in a machine vice. 3. Mount side and face cutter. 4. Set spindle speed and feed rate. 5. Machine ends square and to 100 mm long. 6. Mount slab mill; machine billet to size (43 mm × 20 mm). 7. Turn over and swivel vice through 90°. 8. Mark out slot and tenon on workpiece. 9. Machine tenon underneath. 10. Isolate your machine from electricity supply. 11. Deburr all sharp edges, stamp your name on your work. 12. Clean the work area and machine.	*Horizontal milling machine* *Machine vice and parallel bars* *Side and face milling cutter* *Slab mill* *Slotting cutter* *Hide hammer* *300 mm Rule* *0–25 mm micrometer* *25–50 mm micrometer* *Vernier caliper* *Letter stamps*

List the various types of cutting tools available for use on a horizontal milling machine, stating by ticking the appropriate box whether it is available as:

- HSS
- brazed tip
- replaceable insert.

Note. Some tools are available in more than one type

Tool type	HSS	Brazed tip	Replaceable insert
1.			
2.			
3.			
4.			
5.			

Write here the safety precautions to be taken when operating any type of milling machine:

1.

2.

3.

Write here the safety clothing and equipment to be worn/used when operating milling machines:

1.

2.

3.

Sketch:
Draw a sketch to show how a cutter is secured onto an arbor:

Finished inspection report:

Component dimensions (mm)	Limits:	Actual sizes:	Maximum error:
Width 38	38.25 37.75		
Slot width 22	22.25 22.0		
Tenon width 14	13.8 14.0		
Body thickness 10	10.25 9.75		
Length 100	100.25 99.75		

Results:

Is the workpiece to the drawing specification?

YES	NO

If completed workpiece is below drawing specification, write here the reasons for any errors, stating clearly how the errors will be avoided in future.

Start time and date:	End time and date:	Time taken (hours):

Witness testimony

I confirm that the milling fixture was made within the stated limits and that
. . . . carried out the work in a safe manner and knows why it is important to maintain good housekeeping and equipment control.

Signed . . . . Job title . . . . Date . . . .

Stub arbor

A stub arbor can be fitted to the spindle nose of a horizontal milling machine. Stub arbors are used to hold shell end mills or face mills to machine large flat surfaces. The two principal advantages of using milling cutters in a stub arbor are:

- there is greater machine rigidity
- the workpiece need not be reset between operations.

Stub arbor in a horizontal milling machine

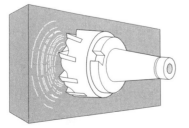

To fit a stub arbor or chuck in a horizontal milling machine

(a) Clean the inside of the machine's spindle and the tapered end of the stub arbor.
(b) Insert the stub arbor's taper into the spindle's tapered bore.
(c) Go to the back of the machine and insert the drawbar into the spindle.

(d) Screw the draw bolt into the end of the stub arbor and tighten with the locknut.

Cutting tools are mounted in the same way as for vertical milling machines (pages 325–27).

Straddle milling

A workpiece can be milled on two surfaces at the same time if **two** side and face cutters are mounted on the arbor of a horizontal milling machine. The spacing collars are selected to give the required distance between the cutters. The procedure is called *straddle milling*. Straddle milling is used when a number of components are required with one feature of the same size.

Procedure for setting up and straddle milling is as follows

(a) Select and mount a work-holding device that will hold the workpiece with the faces to be milled exposed.
(b) Securely mount the workpiece (or a test piece) in the work-holding device, use solid fixed abutments where possible.
(c) Select a matched pair of side and face cutters and a range of spacing collars.
(d) Mount both side and face milling cutters on the machine's arbor separated by collars. The width of the collars should be equal to the required distance between the component's faces.
(e) Carefully measure the distance between the milling cutters' cutting edges. The distance can be adjusted by changing collars for ones with different widths or adding shims. Adjustable milling collars may also be used if available.
(f) Set the machine spindle speed and feed rate.
(g) Position the machine's table so that the cutters cut the workpiece as required.
(h) Take a cut across part of the workpiece.
(i) Isolate the machine's electric supply and measure the workpiece accurately, and if necessary, adjust the collar distance between the cutters on the arbor by either inserting, changing or removing milling collar shims.
(j) Recut the workpiece and if necessary, repeat the above process.

Straddle milling

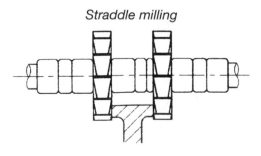

EXERCISE 10.4   Tee nuts

Make four tee nuts using a horizontal milling machine and your milling fixture. Use the procedure and tools indicated. Your finished workpiece should be within the drawing limits. On completion of the task, carefully measure your work and write in the table the actual sizes, noting any errors.

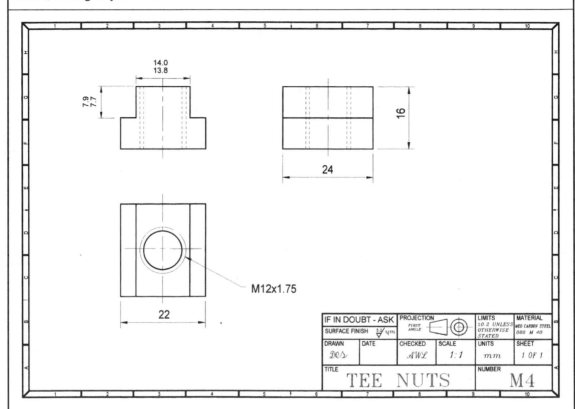

IF IN DOUBT - ASK	PROJECTION		LIMITS	MATERIAL
	FIRST ANGLE		±0.2 UNLESS OTHERWISE STATED	MED CARBON STEEL 080 M 40
SURFACE FINISH 3.2 μm				

DRAWN	DATE	CHECKED	SCALE	UNITS	SHEET
DeS		AWL	1:1	mm	1 OF 1

TITLE		NUMBER
TEE NUTS		M4

M12x1.75

14.0
13.8

7.9
7.7

16

24

22

Planning:

Procedure	Tools and equipment selected
1. Acquire material and check for size. 2. Secure a vice on the machine's table. 3. Hold workpiece horizontal to the milling fixture. 4. Mount slab mill. 5. Set spindle speed and feed rate. 6. Machine top surface to size. 7. Mount two side and face cutters. 8. Machine 14 mm slot to width. 9. Drill and tap M12 holes on a pillar drill. 10. Mount a slitting saw, reset speed and feed. 11. Mount the fixture square in the machine table.	*Horizontal milling machine* *Milling fixture (Job 4.3)* *Slab mill* *Two side and face milling cutters* *Slitting saw* *Pillar drill* *M12 taps* *Tap wrench* *300 mm rule* *0–25 mm micrometer* *Letter stamps*

Procedure (cont.)	*Tools and equipment selected (cont.)*
12. Cut off four pieces to length using the set-up shown in the figure on page 352 of this chapter. 13. Isolate your machine from electricity supply. 14. Clean the work area and machine. 15. Deburr all sharp edges, stamp your name on your work.	

Complete the tables with correct information:

	Diameter	Width	No. of teeth	Bore dia.
Accurate description of slab mill used to machine top of the billet				
Accurate description of the side and face cutter used for the slot				
Accurate description of the slitting saw used to cut tee nuts to length				

Write here the safety **procedures** to be observed when handling horizontal milling cutters:

1.

2.

3.

Write here an accurate description of the machine arbor:

1. Overall length:

2. Diameter:

3. Key width:

4. 'Hand' of thread (left or right)

Sketch:
Make a clear sketch of the
cutter guard used on your
horizontal milling machine.

Finished inspection report:

Component dimensions (mm)	Limits:	Actual size:	Error:
Tee width 14	14.0 13.8		
Overall height 16	16.2 15.8		
Step distance 8 (ref)	7.9 7.7		
Overall length 24	24.2 23.8		
Overall width 22	22.2 21·8		

Results:

Is the workpiece to the drawing specification?   YES   NO

If completed workpiece is below drawing specification, write here the reasons for any errors,
stating clearly how the errors will be avoided in future.

Start time and date:	End time and date:	Time taken (hours):

Witness testimony

I confirm that the set of tee nuts were made within the stated limits and that
. . . . carried out the work in a safe manner and that the trainee
knows the range of operations that the horizontal milling machines are capable of
performing.

Signed . . . . Job title . . . . Date . . . .

Defects on milling cutters

Just like any other cutting tool, milling cutters suffer from wear and must be looked after to get the best performance. Looking after milling cutters means:

- store them individually so that their teeth do not touch each other
- use correct calculated spindle speeds
- use correct feed rates for material being cut
- always use cutting fluid (except for cast iron).

When a cutting tool is not producing a satisfactory surface finish, you should check the following before calling for assistance:

- Spindle speed is correct.
- Feed rate is correct.
- Coolant or cutting fluid is applied.

If the workpiece's surface finish is still poor after the above checks have been made, visually examine the actual cutting edges of the milling cutter – they may be chipped or worn.

If you have been machining soft materials such as aluminium, you will note that these materials are prone to weld themselves to the top face of the cutting edge, which forms a 'built-up edge'. A built-up edge can be avoided by:

Built-up edge

- polishing the front faces of the milling cutter's surface
- using milling cutters with greater rake angles
- careful selection of coolant.

All worn tools should be reported and sharpened for best results. Many companies have set procedures for reporting defective cutting tools.

Other work-holding devices for use on milling machines

To continually achieve accuracy and a good surface finish it is essential that the workpiece is held secure. There are a variety of work-holding devices available to millers. Vices, as we have seen, are quick to set up and convenient to use, but they are not always the most appropriate of the work-holding methods.

Other work-holding devices for use on either horizontal or vertical milling machines are often required and the following factors should be considered when selecting a work-holding device:

- **Workpiece size** – Some workpieces cannot be held in vices because they are simply too big.
- **Workpiece shape** – Irregular shaped workpieces cannot be held in ordinary vices, although special jaws may be made. Vee blocks held in the vices' jaws is a simple way to locate a cylindrical workpiece. If the workpiece being held is sand cast or has a particularly poor surface finish, it would be more difficult to grip securely in a vice.

Cylindrical workpiece

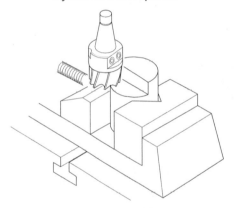

- **Number of faces to be machined in one setting** – A workpiece should be held so that the fewest set-ups are required. This is because setting up the machine takes time and is a situation in which errors can occur. Consequently, when choosing a work-holding device, always make sure you can machine as many surfaces as possible.
- **Quantity of workpieces to be machined** – For batch production, workpieces must be changed quickly, and successive components must locate in the machine in the same position as the last. This is especially important when using unskilled and semi-skilled labour.

Operator safety and workpiece security must always take priority over *all* other factors.

One of the following work-holding devices can be selected depending on the workpiece to be machined:

- **Angle plate** – The workpiece can be clamped to an angle plate when one surface is to be machined square to another on a large component. Angle plates with supporting ribs are best for milling operations as they are most rigid. Bolt the angle plate to the machine table and then secure the workpiece on the angle plate.

- **Clamped direct** – Workpieces can be clamped directly to the machine table. This is often preferred when particularly large workpieces are to be machined.
- **Vee blocks** – Used to mount circular workpieces as shown, to support a workpiece on a machine table.
- **Fixtures** – Useful extensively when successive components are to be machined in the same way, that is, a large batch production. A fixture is expensive to produce but provides consistent workpiece location.
- **Indexing devices** – Used for mounting workpieces that need to be rotated through an angle. Various types of indexing devices are manufactured. A simple indexing unit fitted with a three-jaw self-centring chuck is shown on p. 353. It can be mounted on the machine's table with the workpiece along the table's axis, or with the workpiece's axis vertical.

Large workpiece clamped direct to a horizontal milling machine's table

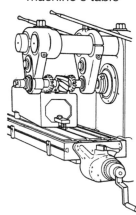

Using angle plate

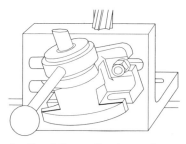

Note the teeth of the cutter force the workpiece against the angle plate due to its direction of rotation.

Courtesy WDS Limited

Clamped on two vee blocks on a vertical milling machine's table

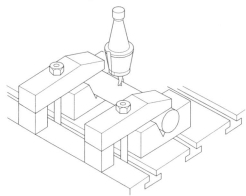

Using a fixture to cut off tee nuts to length

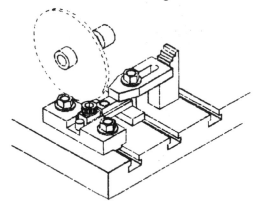

Indexing device

Courtesy WDS Ltd.

Other indexing devices are dividing heads and rotary tables.

Dividing head

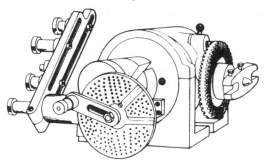

Rotary table

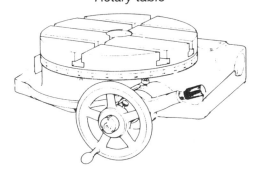

Vertical milling attachment for horizontal milling machines

For versatility, some horizontal milling machines can be converted to a vertical milling machine by fitting a vertical attachment. The vertical milling attachment fits on a horizontal milling machine's column and receives its drive from the spindle nose. Its tools are mounted in the same way as they would be in a vertical milling machine. Vertical milling attachments can be tilted over like the head on a turret mill. This enables the machine to be used in a wide variety of situations.

Vertical milling attachment fitted to a horizontal milling machine

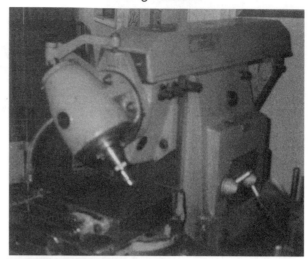

EXERCISE 10.5   Turning tool holder

Make the turning tool holder using either a vertical or a horizontal milling machine. List the appropriate procedure and tools required. Your finished workpiece should be within the drawing limits. On completion of the task, carefully measure your work and write in the table the actual sizes, noting any errors.

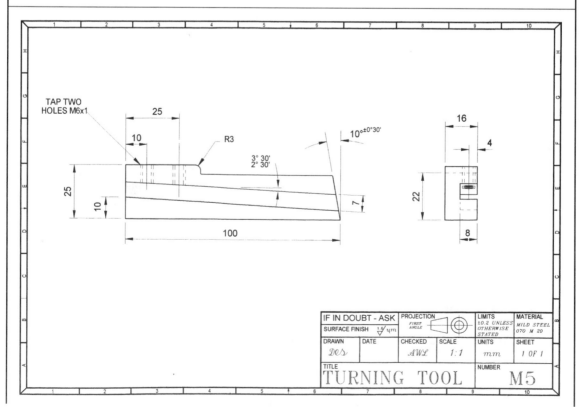

Planning:

Procedure	Tools and equipment selected
1. Acquire material and check for size. 2. 3. 4. 5. 6. 7. 8. 9. 10. 11. Isolate machine from electric supply. 12. Clean the machine and the work area.	

Complete the tables with correct information:

	Reasons for selection
Machine type selected:	1. 2.
Work holding device selected:	1. 2.
Cutting tools selected:	1.
1. 3.	2.
2. 4.	3.

Write here the checks made on cutting tools before starting work:

1.

2.

3.

Write here the procedure for reporting defective cutters in your workplace:

1.

2.

Sketch:

Draw two sketches to show alternative methods of how the 10° end of the turning tool holder could be held for machining in two different work-holding devices.

Finished inspection report:

Component dimensions (mm)	Limits: (trainee to complete)	Actual size:	Error:
Overall height 25			
Slot width 7			
Body height 22			
Slot depth 8			
Slot angle 3°			

Results:

Is the workpiece to the drawing specification?

YES	NO

If completed workpiece is below drawing specification, write here the reasons for any errors, stating clearly how the errors will be avoided in future.

Start time and date:	End time and date:	Time taken (hours):

Witness testimony

I confirm that the turning tool holder was made within the stated limits and that. . . . . .carried out the work in a safe manner and cleared the work area on completion of the task.

Signed . . . . .Job title . . . . Date . . . .

11 Using computer software packages to assist engineering activities

Exercise checklist

(Ask your assessor to initial the box for each exercise when you have completed it.)

Exercise No.	11.1	11.2	11.3	11.4	11.5	11.6	11.7	11.8
Initials								
Date								
Exercise No.	11.9	11.10	11.11	11.12	11.13	11.14	11.15	11.16
Initials								
Date								

The use of computers in engineering is common. Computer numerical control (CNC) machines and computer-assisted design (CAD) are two such uses. The size and complexity of the computer determines the extent of the work that can be done and the software determines the nature of the work.

While working through this Chapter, you should gain most of the evidence you need for information technology (IT) key skills at Level 2. You do not need access to very specialised software but by the end of this unit you are expected to show that you can use software packages that assist in engineering operations.

Very importantly, you need to operate computers and demonstrate that you are doing this competently and safely. It is not enough, in this case, to do some exercises in this book; you need to get your supervisor to sign that you have been seen carrying out a number of activities. If you do not do this, you will neither get this unit nor will you be accredited with IT key skills.

Working safely with computers and peripheral hardware

Before considering details about software and hardware, it is important to think about health and safety issues. People spend a lot of time sitting in front of computer screens and this brings forth issues that should be considered. The main problems associated with working for extended periods at a computer workstation are as follows:

- Musculoskeletal problems. The main problems here are repetitive strain injury and stress to the back and upper limbs from sitting poorly.
- Visual tiredness and eye strain. The main problem here is tired eyes due to extended time at the workstation. Actual eye disease or permanent damage is not very likely.

- In association with tired eyes, thought needs to be given to the quality of the visual display unit (VDU), particularly screen glare and image quality.
- A person may also need to use different spectacles when working with a VDU.
- A problem for some people is that they become anxious while working with new technology and this stress makes them more tense and adds to the likelihood of musculoskeletal problems.
- Many people tend to spend long periods of time in front of the screen and this, coupled with poor posture, can lead to problems that could easily be alleviated by taking regular short breaks away from the screen.

Notice the open window, the footrest, the protected leads, the screen at 90° to the windows that have blinds in front and the swivel chair on castors. She is not sitting very well but her face is at the right height for the VDU.

If you look at the drawing above, you can see that there are other potential hazards as well.

- Note that the trailing cables have to be covered with rubber guards.
- Some people exhibit an allergy to screen cleaners (although this is rare).
- Lighting should be adequate for both VDU and non-VDU use. Between 300 and 500 lux is recommended.
- Noise from cooling fans and extractor fans plus keyboard tapping and people talking can be stressful for some people and cause fatigue. Noise at a workstation should not exceed 55–60 dB (decibels).

The best way to avoid problems is to think about workstation design and work activity. If necessary, a risk assessment on the workstation could be carried out. However, try to avoid screen glare and reflection by positioning the monitor at right angles to major sources of light and cover windows with adjustable blinds if necessary. Ensure that the chair being used is stable. The best type of chairs are swivel chairs with a five-star base on castors. Such a chair is stable and the swivel action prevents someone from twisting the body awkwardly. The height of the seat should allow the person to rest his/her feet comfortably on the floor, or alternatively, a footrest should be provided.

Chair adjustments from Global Supplies catalogue

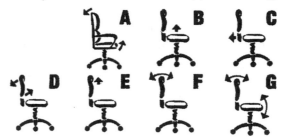

The desk should be wide enough to allow 100-mm space in front of the keyboard so that the user can rest his/her wrists on the desk while using the computer.

You should be instructed on the correct use of a computer workstation, and supervisors or managers may ask you to adjust your working position in order to avoid long-term problems.

A few other dos and don'ts with regard to working safely with computers:

Do:

- Look after your eyes. There are VDU regulations that govern the use of VDUs, and people who are required to spend a lot of their time looking at VDU screens should get their eyes checked and this should be paid for by your company.
- Check that there are no loose wires and connections.
- Make sure that the chair you use conforms to modern safety regulations. The study of workstation design is called *ergonomics* and a badly designed workstation can lead to health problems.
- Think about your health and sit properly at the correct height to avoid back and repetitive strain injury problems.
- Make sure that you know the correct procedures for reporting faults and breakdowns.

Do not:

- Take food and drinks near computers. Liquids and electricity do not mix.
- Attempt to fix a problem if you are not entirely sure of what you are doing. It is one matter to fiddle with your own computer; it's quite another problem to cause an expensive system to crash as a result of your meddling.
- Annoy others by 'crashing' a system.

EXERCISE 11.1 Working safely with computers

1. Identify three risks associated with working with computers.	1. 2. 3.
2. For each risk identified, describe how the risk could be minimised.	1. 2. 3.
3. Which overall Act of Parliament governs the use of computers by people in a workplace?	

4. Which other regulations need to be complied with?
 	
  (a) COSHH
  (b) VDU
  (c) Manual handling operations
  (d) Personal protective equipment regulations	

The above exercise has been completed satisfactorily by.....................

Signed................Job title .... Date..........

Working with computers

While talking about computers, we often come across two terms, 'software' and 'hardware'. Hardware is the computer equipment, the computer itself, the keyboard, mouse, VDU screen, printer etc. while the software is the electronic instructions that tells the computer what to do.

There are two main types of computers – personal computers (PCs), designed to be used by individuals, and mainframes. Mainframes can process and store vast amounts of information. Many organisations use both types depending on the nature of the work to be done.

A basic computer network

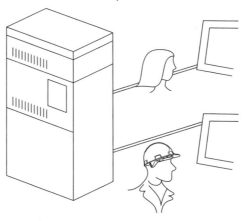

A computer

Peripheral hardware is the equipment that supports computer activities. Peripherals that are readily available include the keyboard and mouse, the VDU, the printer, scanners and plotters, and modems. As the hardware needs to be powered by electricity, it is essential to maintain this equipment in safe working order.

When a PC is set up for the first time, the various pieces of hardware need to be connected correctly. Connections are made using different types of cables. Cables connect computers and equipment to each other and to a network. The most common is the coaxial, which is the same type of cable that is used to connect your TV to the aerial. If you look at coaxial cable, you will see that there is a solid internal core of copper with an outer mesh in another type of metal. You will also find that there is an unshielded twisted pair (UTP) cable and a shielded twisted pair (STP) cable. Finally, for rapid communication of information over long distances, there is a fibre optic cable. This type of cable is expensive but does a very good job.

The actual nature of the cables being connected is often difficult to determine as it is covered in insulation and specialised connectors are already in place. These specialised connectors can make connecting up the computer and peripherals almost foolproof. Connectors fit into ports. A connector is basically the male end of a connection and a port is the female end. Look at the drawings below to see the different types of connector and ports you may come across. There are five main types of port. A **parallel port** will have 25 holes and may be used to connect a printer to a computer. There are also **monitor ports**, **serial ports**, into which the mouse or a modem are connected, a **keyboard port**, and a **joystick port**.

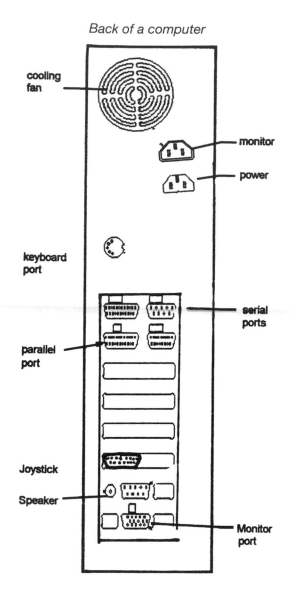

Back of a computer

It is essential that these connections are made correctly because otherwise it could be damaged and the equipment can be made unusable. Therefore, if you come across cables that have become dislodged and equipment that is not working properly, you must report the matter and not risk making wrong connections.

EXERCISE 11.2 Natural performance evidence		
You must be **seen** carrying out the following operations on **two** occasions:	Date	Assessor Initials
Checking that all equipment (VDU, monitor, keyboard, mouse, printer) is correctly connected and in safe working order, with any faults noted and reported		

Powering up equipment using the correct procedures and checking that each item of equipment is functioning correctly		
Using the correct log-off and shut-down procedures		
Identifying and discussing any problems or difficulties with your supervisor and carrying out the actions agreed		
Disposing of waste paper or surplus materials in line with organisational procedures and maintaining the work area in a safe and tidy condition		

I confirm that. . . . . has been seen to carry out all of the above criteria satisfactorily.

Signed . . . . Job title . . . . Date . . . .

Different types of display screen equipment

Your computer is connected to the monitor by a **video card.** A video card is a circuit board and its job is to translate the information from the computer into words on the screen or pictures. Video cards may also be called *graphics cards*.

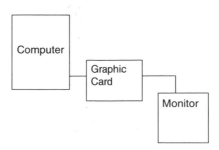

Another type of screen is found in portable computers. Ordinary monitors are heavy, whereas portable computers or **laptops** use a liquid crystal display (LCD) that is much lighter. In order to see this display, there is an internal light provided behind the LCD screen. This light is called a ***backlight***.

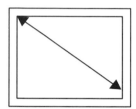

You may have noticed that laptops vary considerably in price and one reason for this is the type of LCD screen incorporated. Cheaper laptops probably have a passive matrix that is less bright and is difficult to see from an angle. The one benefit of such a screen is that it gives the operator some privacy. An active matrix screen provides a brighter image and can be useful for presentations as it can be viewed from a wider angle. It is also possible to connect a computer to a large television screen to allow presentations to much larger numbers of people.

All screens are measured diagonally. You should note that the manufacturer will measure the screen size for monitors as the diagonal distance across the tube, and this means that the screen size stated may be actually greater than your viewing area, as part of the screen will be covered by the casing. We still use inches to measure screen size, and the larger the size, the greater the price. However, for engineering drawings and design work, the larger screen is almost essential.

If you are going to work at a screen for a long time, there are other things apart from screen size that can make life more comfortable. Crisp images are less tiring to look at and the crispness is determined by the **dot pitch** that is actually the distance between the dots on the screen. A dot pitch of less than 0.28 mm provides a good crisp image.

Screens that flicker can also be very tiring. Modern technology gives us the **non-interlaced** monitor with less flicker. Another cause of flicker is that the monitor needs to **refresh** or update images on the screen. High refresh rates of 72 Hz (Hertz) are better for your eyes.

Most monitors can also be adjusted for brightness and contrast. There is often a control panel below the screen or on one side of the monitor.

Another term you may come across with regard to monitors is **resolution**. If you could look at a screen under a microscope, you would see that it was made up of thousands of rectangular units like a graph paper. These units are the smallest units of the screen and are called *pixels*. The number of vertical and horizontal pixels determines the resolution. Basically, the more pixels there are, the more information you can see on the screen.

Magnifying glass in front of the VDU showing pixels

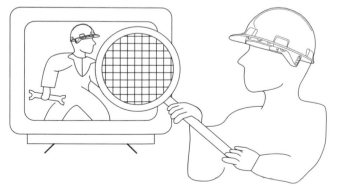

Finally, we notice that some monitors give us better depth of colour than others. Most modern computer systems operate with **super video**

graphics array (SVGA). Totally realistic colour is provided by 24-bit colour. In this set-up, the computer can generate 16 777 216 colours, which is actually more than a human eye can distinguish! 16-bit colour provides us with 65 536 colours, which is enough for most people.

We tend to accept what is in front of us but we can make life more comfortable and safe for ourselves by being aware of the following:

Be aware:

- That you can usually adjust the tilt of the screen.
- That you can often swivel a screen so that it does not reflect the glare from other lights.
- That you can also obtain a glare filter that reduces reflected light.
- That there are EU regulations governing electromagnetic radiation sources. Monitors emit em radiation and there is more from the sides and back than from the front, and the further you are away from the screen, the lower the dose.
- That old monitors generated a lot of static energy and this was enough to cause oxygen in the atmosphere to be converted to ozone. Low-level ozone can cause headaches and dizziness and large amounts of it contribute to urban smog and pollution.
- That you can help protect the environment by making sure that the equipment is turned off when not in use.

Using peripherals

Printers are the hardware that convert the electronic information in your computer to a paper copy. Printers range considerably in price and performance. Dot matrix printers are the cheapest, and dry sublimation printers that produce pictures that look like colour photos are the most expensive. Companies and individuals will purchase printers that meet up to the tasks required of them.

Printers

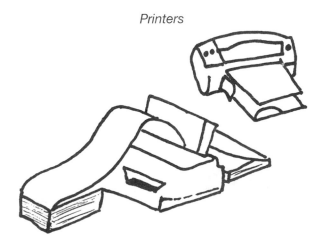

Printing speed is usually very important in a busy organisation. Laser printers are the fastest printers. **Resolution** is a measure of the sharpness of the image produced. Resolution is measured in dots per inch (dpi). The greater the resolution, the sharper the image. For most applications, 300 dpi is satisfactory but many modern printers offer 600 dpi with 1200 dpi being available, but at a greater cost.

Colour printing is now regarded as almost essential. The trouble with colour printing is that most of the printing will be in black with the colour used less often. Some printers have exchangeable cartridges so that when colour is needed, the colour cartridge is slotted into place after the black cartridge has been removed. Alternatively, the colour option is always available but the problem is that the black ink may run out quite quickly and so refillable cartridges may be useful. Colour laser printers are now available but are expensive.

For engineering drawing purposes, printers that will accept larger paper sizes are useful, for example, A3 instead of A4. However, for office work in an engineering company, a printer that will also print envelopes and labels is essential. Finally, in a busy situation, the printer needs to carry a lot of paper, and in the case of the paper jamming, simple procedures for unjamming are desirable.

As with all equipment, people need to know how to operate it, and if there is a problem, they should only attempt to correct the situation if they know how to. Amateur meddling with office equipment when a service engineer needs to be called in can cause real problems and involve a company in extra expense.

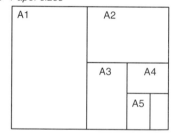

A0 *Paper sizes*

O.K. That's fixed now. People shouldn't meddle!

Types of printer

- **Dot matrix printers** – These are less widely used than they used to be. They work by causing a printhead with blunt pins on it to strike the paper through a ribbon rather like a typewriter's ribbon. Most dot matrix printers use continuous paper which is paper pleated into a large stack. The sides of the paper have holes punched that feed the paper through the printer. One major advantage of this type of paper and carriage is that the paper can be in different sizes and can print large spreadsheets. A common use for dot matrix printers is issuing invoices. The reason for this is that because the image is produced by impact, a second copy can be formed – one for the customer and one for the company. These printers are noisy!

- **Inkjet printers** – These produce an image by spraying ink on the paper. They are ideal for producing routine documents and materials. Their main problem is their speed with a maximum output of about 4 pages per minute. Amazingly, only three colours, in addition to black, are used to produce the vast array of colours that can be produced. The three colours are magenta (reddish), yellow and cyan (bluish).

- **Laser printers** – These are high-speed, high-quality printers that work like photocopiers. The ink in a laser copier is a powder called *toner* that comes in a cartridge. Cartridges are quite large and do not need replacing too often. Laser printers can work quickly, printing up to 20 pages per minute with good resolution. Colour laser printers are also available and their images are superior to those produced by an inkjet printer. The main drawback to colour laser printers is their cost. In order to produce the quality of image possible in laser printing, the printer has a memory, typically between 1 and 4 megabytes. This memory is essential because the computer can send information to the printer faster than the printer can take action on the information. This is particularly true when high-resolution printing is being carried out.

Another reason for memory in a printer is that the printer may be connected in a general network set-up and several people will be sending work to the same printer. Slowness of printing may be due to a long 'print-queue'.

Other types of printer include **multifunction printers** that may also incorporate fax, scanner and photocopier facilities. **Solid ink printers** are used to produce high-quality colour images and are very useful for making overhead transparencies while the very expensive **dry sublimation printers** produce a quality to equal colour photos.

Different departments in an engineering company may have different printer facilities and some printers may be connected to computers with special software.

Plotters are another type of output device and are often used with CAD software. The ink cartridge is connected to a number of pens and information is fed to each pen individually. In addition to the pens moving and drawing, the paper in the plotter also moves from side to side as well as through the plotter. Plotters can take very large sheets of paper, up to size A0, so that very large drawings can be made.

Scanners are input devices. They are able to 'read' both images and text and feed this information into the computer and on the screen. Scanners are extremely useful and provide a quick method of copying information into computer formats that can then be stored, printed or sent elsewhere. Scanners are a type of hardware but they need specialised software; the most common one allows editing of the scanned image. Another aspect of the software is **optical character recognition (OCR)**, which means that the scanned text can be used with word processor software.

Scanner

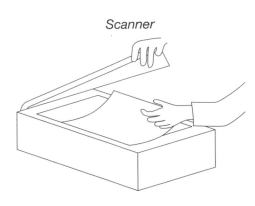

As with printers, there are different types of scanners that have different uses. Probably the most useful is the **flatbed** scanner. In this type of scanner, the image is laid flat on a sheet of glass that forms the bed of the scanner and a lid is brought down to keep out incidental light. An open page within a book can be scanned in this way. Other types of scanner are **hand-held scanners and short-fed scanners**. Short-fed scanners can scan single pages, so a page would need to be removed from a book in this case, and hand-held scanners are pushed manually over the image. These are very good for scanning small amounts of information but are very slow for large images.

As with printers, the quality of the image depends on the resolution. The higher the resolution, the more information is being fed into the computer and the slower the operation. Resolution is also measured in dots per inch (dpi).

Like printers, scanners can also work in colour or in tones of grey. Greyscale scanners are ideal for scanning drawings and will produce accurate images with a black and white printer.

Modems are the hardware that allows computers to communicate via telephone lines. For personal use, most people do not have a separate phone line for their computers. When someone is 'on-line', the phone line is engaged. Companies may have a dedicated line and, increasingly, they may have an ISDN line. ISDN stands for Integrated Services Digital Network and these provide high-speed connections. ISDN replaces the need for a modem and works about four times quicker than a modem. This facility is of worth if information has to be sent long distances and overseas, as the amount of time 'on-line' is much reduced and telephone charges will be lower.

There are two main types of modem. Internal modems are located within the computer itself, whereas external modems plug into the back of the computer. The main advantage of the external modem is that it can be networked to more than one computer.

The **speed** at which a modem works is important. Speed is measured in kilobits per second and speeds of less than 14 400 Kbps seem slow. Anyone who has ever sent an e-mail will appreciate how fast modems can work. The only problem with speed is that if your modem is connected to a modem that is slower than yours, then the speed will be at the speed of the slower modem. When connections are made, modems perform a **handshake**, which is the term used to explain that modems need to establish how they will exchange information.

Speed can also be increased by compressing data. Data compression is easier for text files than for graphics files. An example of file data compression is the use of **.jpg** as a file extension. If you are used to using a word processor, you will know that when you save a document in Word, for example, the computer will add **.doc** to the file name. This is an example of a file extension and .jpg is another one. If you use a scanner and want to save an image, you may be prompted to choose that file extension you want to use, and the use of .jpg is useful if you want to compress the data for transfer to another computer using a modem.

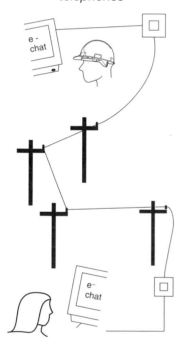

Modems connected via telephones

EXERCISE 11.3 True or false?		
Statement – Tick the true or false box to answer the question.	*True*	*False*
1. The electronic instructions given to a computer are termed 'software'.		
2. A computer is attached to the monitor via a video machine.		
3. Standard computer monitors use a liquid crystal display (LCD).		
4. The smallest unit on screen is called a pixel.		
5. Laser printers are the fastest form of printer.		
6. Resolution is measured in dots per inch (dpi).		
7. High quality, expensive printers contain up to 10 colour cartridges.		
8. Computers with CAD software are often connected to a plotter.		
9. ISDN stands for Integrated Specialised Design Numbers.		
10. Data compression speeds up the process of transmitting information using modems.		
The above exercise has been completed satisfactorily by............................		
Signed . . . . Job title . . . . Date		

Input devices

Input devices allow you to communicate to the computer. The basic input device is the keyboard and all operations can be carried out using a keyboard only. Sometimes keyboards are referred to as *QWERTY boards* after the first six letters on the board.

However, as computers became more sophisticated, the mouse was developed. The mouse allows you to navigate around the screen very quickly. It is useful to know that for left-handed people the mouse controls can easily be reversed. A mouse is usually operated using a mouse pad. The reason for this is that it gives the mouse a clean, smooth surface on which to operate and prevents the mouse from becoming too dirty.

A mouse can easily be cleaned by removing the protective cover on the under surface and cleaning the mouse ball. It is possible nowadays to have a cordless mouse that sends messages to the computer like a remote control for the television.

Different types of mouse

A **joystick** is another input device that allows for very rapid movement around the screen but it is used mainly for computer games.

Touchpads are very useful for people using graphics software. The surface of the touchpad, is sensitive so that when you move your finger across the touchpad, the pointer on the screen also moves.

Trackballs are often incorporated into laptops. They are basically an upside down mouse in which the trackball remains stationary and you move the ball, often with your thumb. Variations on this idea help people with limited movement in their hands. The size of the ball can be made quite large to make movement easier. Similarly, touch screens can be adapted for use by people who are paralysed.

Other input devices that are much in use are scanners, including bar code scanners, and digital cameras.

If the keyboard or any other input device is not connected properly, your computer screen will show a message during the start-up procedure. It may tell you to connect the keyboard and then press F1.

Start-up and shut-down procedures

When a computer is first turned on, a great deal of information flashes on the screen. If there are problems with the start-up procedure, an IT engineer can look carefully at this information to sort out where the problem lies.

With start-up procedures, there is very little an individual user can do to speed up the sequence of events. With **shut-down** procedures, it is important to allow the computer to shut down properly. Just as an athlete needs to warm up and warm down after competing to avoid injury, a computer needs to be turned on and turned off in the proper way. If the proper procedures are not followed for shut-down, the next time the computer is switched on, a message will appear on the screen saying that the computer was not shut down properly and that the user must 'press any key to continue'. The start-up procedure will then include scanning software for faults and viruses and will take longer than normal.

Where there is a computer network, access to the network may involve each user entering their individual password. If this person does not **log off** properly, the next person may have difficulty in gaining access to the software and being able to operate the computer. It is essential that all users follow the correct logging-off procedures as it can seriously inconvenience others if they do not do so.

It is essential that you know the correct procedures for the systems that you operate and you will notice that Exercise 11.2 requires that you are seen to follow these procedures.

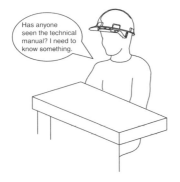

Technical manuals

Each computer and peripheral will come supplied with a technical manual. If you are in any doubt as to what type of connectors are needed to connect the peripheral to the computer you are using, then the information will be in the technical manual. This information is most important if a decision to buy a new peripheral has been made and you are asked to write out the order. You will need to check on the exact nature of the cables and connectors required, for example, whether you need a five-pin or a six-pin keyboard connector.

Good housekeeping

This term refers to keeping the work area tidy. The main problem with computers is waste paper. Waste paper comes from empty wrapping paper from new reams of paper and from unwanted printed sheets. Waste is expensive. Most firms try to recycle waste to avoid sending it to land-fill. Waste paper bins are provided for this purpose but the obvious thing is to try to cut down on too much waste. Work should be checked on-screen before being sent to the printer. Many companies are now working towards the paperless office. This means that all information is sent electronically via internal e-mail. Good housekeeping means being responsible, keeping the work area tidy and clearing away any mess you make.

EXERCISE 11.4   Complete the following multiple choice test by circling a letter from A to D

1. The following regulations apply to the use of computers in the workplace: A. COSHH B. Eyeshield Regulations (1974) C. Provision and Use of Work Equipment Regulations (1992) D. The Electricity at Work Regulations (1989)	2. Potential hazards in the workplace include: (i) a cluttered work station (ii) trailing leads (iii) dripping taps (iv) frayed wires A All of the above B (i), (ii) and (iv) C (i) and (iii) D None of the above.
3. It is important to keep the workplace clear because: (i) lots of loose papers mean that documents are easily lost (ii) paper is a fire hazard (iii) people can slip on a piece of paper on the floor (iv) mice make nests in computer paper A All of the above B (iii) and (iv) C (i), (ii), and (iii) D None of the above.	4. When you turn on the computer, a message flashes on the screen that it was improperly shut down. You should: A follow on-screen instructions B rewire the mouse C switch off the computer and switch on again D reboot the system
5. You send a document to the printer but nothing happens, so you: (i) click the Print command five times for luck (ii) assume someone else will sort out the problem	6. You have completed your work on the computer and your work is still on the screen. You have made a single copy of the work. The correct procedure is to: (i) leave everything as it is and go to lunch

(iii) follow on-screen instructions and
check the status of the printer
(iv) check that the printer has paper

A All of the above
B (i) and (ii)
C (iii) and (iv)
D (i) and (iv)

(ii) assume that the single copy is
sufficient and switch off the computer
(iii) save your work, store it in the correct
folder, follow the correct shut-down
procedure and log off
(iv) assume that the single copy is
sufficient and log off following the
correct procedure.

A (i)
B (i) and (iv)
C (iii)
D (ii) and (iv)

7. Your firm has purchased a disk writer (this
will save on a CD instead of a floppy disk)
and you have been asked to set it up.
(i) You look carefully at all the
components and leads, read the
instructions and then proceed. You
make sure that there are no 'live'
connections before you start.
(ii) You read up all about the different
types of disk writers.
(iii) You take a lead from the box, look at
the connector and find a suitable port
in the computer. You plug in the lead,
attach the disk writer and switch on
the computer.
(iv) You delegate the task to your mate
who is a 'whizz' with computers.

A (i)
B (ii) and (iii)
C (iii) and (iv)
D All of the above

8. As a result of a drink being spilt on a
keyboard, there is a computer, which has
specialist software on it, without a
keyboard and you are instructed to
connect up a spare keyboard. The
keyboard has no connecting lead.
(i) You look carefully at the two sockets
and go in search of a lead
(ii) You consult the technical manual to
see what type of lead is needed
(iii) You remove the lead from another
computer and squeeze it to make it fit
(iv) Because there is no lead you tell the
supervisor that the job cannot be
done

A (i) and (iv)
B (ii)
C (iii)
D All of the above

The above exercise has been completed satisfactorily by...............................

Signed.................Job title...............................Date..........

EXERCISE 11.5   Find eight technical terms in the following word search and match them with the definitions below

```
N   M   R   S   C   W   T   A   P
M   C   P   L   O   T   T   E   R
E   O   T   L   M   R   D   J   E
D   S   N   O   P   M   P   O   T
O   H   U   I   U   O   Y   Y   N
M   S   R   S   T   D   U   S   I
E   V   I   P   E   O   U   T   R
B   C   A   K   R   M   R   I   P
T   L   A   S   D   K   I   C   R
D   R   A   O   B   Y   E   K   P
```

1. An electronic processing unit

2. Allows one computer to communicate with another via a telephone wire

3. Colloquial name for a portable computer

4. Allows very rapid movement around screen, often used for games

5. A very common output device

6. Another output device commonly used with CAD software

7. Also known as a QWERTY board

8. Not all of them have tails nowadays! (Not very technical)

The above exercise has been completed satisfactorily by............................

Signed.................Job title...................................Date..........

Working with computers

In this section, you need to show that you can work in an organised way with software. You need to demonstrate your understanding of file management and document control systems, and to do this you have to collect evidence from someone watching you or by printing documents that have been created by you.

EXERCISE 11.6 (See Exercise 11.8 for suitable tasks)		
The things you must prove that you can do	*Witnessed by*	*Alternative evidence assessed by:*
1. Create directories and file structures that allow efficient file storage.		
2. Use simple file management systems and functions to save, transfer and delete files.		
3. Correctly back up, label and store the files.		
4. Ensure that your arrangements for storage of the backed-up files are safe and secure.		

This confirms that............................has been assessed as being able to carry out all the above to a satisfactory standard on two occasions.

Signed................Job title...............................Date..........

Being organised about working with computers

When you use the computer to produce a piece of work, a drawing, some script or some figures, you can either just print the work or you can store it. If you choose the latter, then the stored document is called a *file*. You must now choose a name for this file. Remember that over the days and months, you may create a lot of files, and so you need to think of sensible ways of naming them so that they can be found again quickly and efficiently. In an office, it is part of a secretary's job to file all the information that comes into the office. This person will have a filing system.

In a doctor's surgery, the obvious way to file information is under a person's name and all the names filed in alphabetical order. The doctor may also want to know who is using various medicines, so, in this case, a file entry may be made under the name of the medicine, again in an alphabetical order.

All offices have some sort of filing system and the underlying reason for this tidy organisation is that information can be found quickly and efficiently. This is the principle you need to apply to naming and looking after your files.

Whatever software you or your company use, there will be some method of naming and storing files. Also, you may have your own floppy disk or CD on which you store your training files. Most people

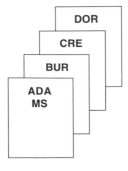

today use 3.5 inch floppy disks, although there are still a few 5.25 inch floppy disks in use. (These are really quite floppy.) We will imagine that your first piece of work is a CV and you need to store it. You finish the piece of work and save it. The obvious name is CV but then you update it. So now what do you call it? Perhaps the new CV could be called CVDecember01, which indicates the date when it was updated. By this time you are saving a lot of information.

You decide to become organised! You buy a box of disks rather than one at a time and you have one disk for engineering drawings and another for reports and another for personal information. You put a sticker on the front of the disk to tell you the contents of the disk. What you are doing is developing your own filing system.

You can develop the same storage system inside your computer on the hard drive. Different companies will have different systems and you need to be shown how the system in your company works. Where there are lots of users, each person will have his/her own name and password and he/she will have to **log-on** to the system. The password is for security and you should not tell anyone what your password is.

Well labelled disks

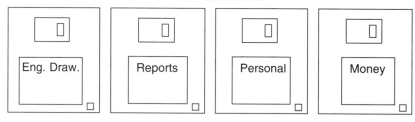

Eng. Draw. | Reports | Personal | Money

Effect of magnets on a computer disk

Once you are logged on, you can access your own files within the computer you are using. The part of the computer that is used for storing information is called the *hard drive*. For a small home computer, there will probably be only one hard drive and this is called *Drive C*. In large systems, there may be lots of drives and these are named D, E, F and so on. All the information that you stored is stored magnetically on rotating disks or **platters**. You need to be shown how your system in your company works. In some cases, just logging on will take you to the correct drive; in less complex systems, you may need to choose the drive you need. If you are storing information on a floppy disk, you will choose drive A for 3.5 in. floppies. Drive B may not exist. Some computers use Drive B to access the old 5.25 in. floppy disks.

Because information is stored magnetically, it is essential that disks are stored away from energy sources, such as magnets or coils of wire carrying electric currents, as these can corrupt the stored information.

Creating directories and file structures

When you become an authorised user for your company's computer system, you will be told which drive you must store your files in (for home computers, this will be Drive C). Once again, you need to learn to become organised and not end up with a system like the one below.

File Contents:
Bona
CV
CVDecember01
Doc.1.doc
Gaz
Edz
Hols
Mum

As you can see most of these may be difficult to find again.

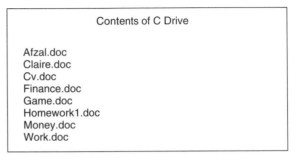

Contents of C Drive

Afzal.doc
Claire.doc
Cv.doc
Finance.doc
Game.doc
Homework1.doc
Money.doc
Work.doc

Now imagine an office in which all the files in the office were kept in one very large box. They might be in alphabetical order but are in a muddle; we have a file for 'July invoices 99' next to 'July orders 98' and they both come after 'Jones', who happens to be a customer, and just before 'June orders 97'. It would be much more sensible to put all the invoices together, all the customers together and so on. You can get your computer to do this for you with your files. It is called *creating files, folders and directories*.

It is like having electronic filing cabinets. Each drawer of the cabinet is named and inside each drawer there are files organised in order. Look at the information that follows, which gives one example of a file system within a folder.

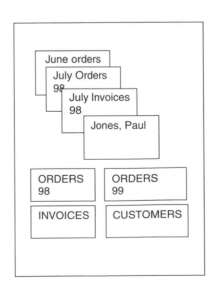

EXERCISE 11.7   Sort the following files into three different categories (folders) and name each category (folder)

Workbox drawing, Claire letter, Design 1, application letter to U_PAK, paperclip drawing, homework for Alan, maths homework, birthday card design, sketches, letter to Claire.

Folder name:	*Folder name:*	*Folder name:*

The above exercise has been completed satisfactorily by..................................

Signed . . . . Job title . . . . Date . . . .

Instead of keeping everything together inside the computer, it is possible to sort files out into folders as can be seen below:

How do you do this? Once again, the best way is to learn from experience and follow on-screen hints.

Here is how to do it:

Click on **File** and when the scroll down menu appears, click on **Save As.** Depending on your system, one of several things might appear at this point.

When **Save As** was clicked, the information was stored in **engbook**, which is the 'filing cabinet drawer' in a file called **Unit 10.2**, which makes sense, and can be quickly found the next time this file is needed. Notice that down the left-hand side there is more choice, for example, History, My Documents, Desktop etc. These big categories are like the separate filing cabinets.

(Alternatively, some people use the various icons at the top of the screen instead of using the File menu. You should use whichever system you are happy with.)

When writing this book and this Chapter, the file name was Unit 10. Naming a file is easy – just make sure that the name is logical and if one file follows another, give it a number and date as well. When you have finished the work, you 'click' on File and a menu rolls down that gives you the prompt: **Save as.** When you click on **Save as**, another menu box appears that allows you to save your file with a name. The nature of the menu will differ according to the software installed in the computer you are using.

Once you have a filing system set-up, you need to keep it tidy. Files take up memory space on your hard drive and, although extra memory can be added to computers, it is a good idea to check through your old files and get rid of any you do not want any more. This is called *deleting* and there are several ways to do this. Look at the screen printed below:

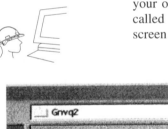

Opening a file

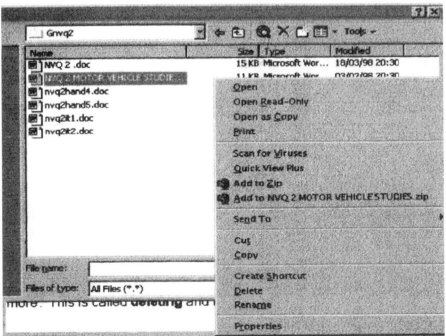

First of all, the folder Gnvq2 was opened and the file NVQ Motor Vehicle Studies was chosen by clicking on it. To obtain the menu shown, the **right mouse button** was used. If you look near the bottom of the menu, you can see <u>**Delete.**</u> To delete the file, all you need to do is click on delete with the left mouse button and a box will appear asking you if you want to send the file to the recycle bin. By clicking on YES, the file will disappear. It can be retrieved by going into the Recycle Bin. Files in the Recycle Bin are deleted only if you empty the Recycle Bin. In fact, this is the same as throwing a paper file into a waste paper bin and then being able to retrieve it until the waste is taken away.

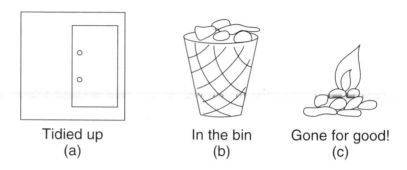

Tidied up	In the bin	Gone for good!
(a)	(b)	(c)

Note that you can do a lot of other things with the file as well. It can be sent to another folder, copied, renamed and so forth. To carry out any of these actions, all you have to do is to click on the action on the menu bar.

Backing up files

When you save files on hard disk, it is possible to lose these files and it is a good idea to have a back-up system for really important information. There are various ways that you can do this. The simplest way is to save on a floppy disk. To save on a floppy disk directly, you have to indicate to the computer what you want to do when you click on **Save As**.

In order to save a file on another drive (Drive A = floppy disk), you click on the file you want to save – in this case NVQ2.doc – and then click again with the **right** mouse button. This provides a menu and Send To is selected. You can see the various options available. The top option is 3 1/4 Floppy (A), but you could also send this file via e-mail using Mail Recipient.

Shows the scroll down menus

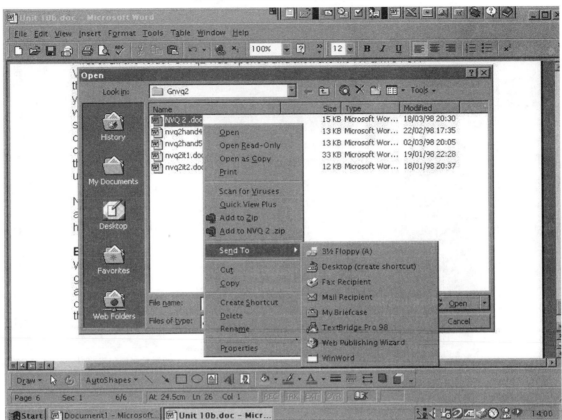

Another way to save information, which many firms do on a regular basis, is to have a computer with a removable hard drive. Zip and Jaz drives are popular forms of removable hard drive. Very sensitive data may be removed every night and placed in a safe. When files are updated at the beginning of a new year, old files may be downloaded on CDs (or disks) or removable memory and stored as archive files. CDs, floppy disks and removable memory are useful ways of transporting data from one place to another. Increasingly, e-mail is also used to do this but security may be a problem here.

In engineering companies, the sort of information that may be vital and that need to be backed-up would be drawings. Archive data here would be old drawings. Removing them from permanent storage on the computer, where they take up a lot of memory, makes sense. Files in store need to be kept away from magnets and extremes of hot and cold. Most firms have special safes that are designed to withstand fire and explosions, and this is where they keep their really important data.

Once you have learnt to name files and save data into designated folders, you also need to learn how to set up new folders. Look at the screen printed below:

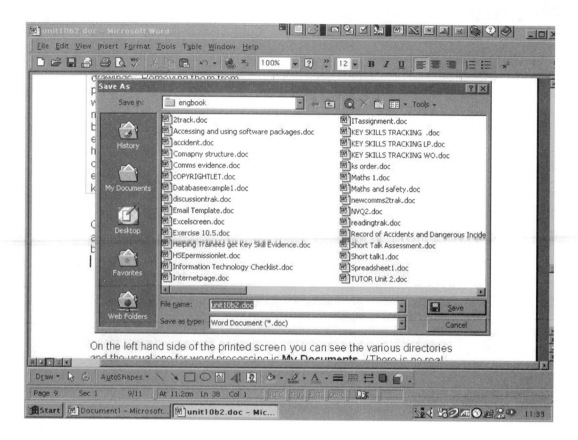

On the left-hand side of the printed screen, you can see the various directories and the usual one for word processing is **My Documents.** (There is no real difference between a folder and a 'directory' and the term directory is not much used nowadays.)

If there are a lot of authorised users for a system, you will have your own folder that is identified by your name. Once you are working on the system, you can create new folders that are stored in your (large) folder.

Create new folder

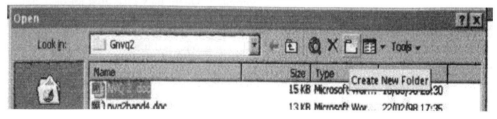

Look at the figure above and you can see that creating a new folder is easy. You click on the Create New Folder icon and then a box appears and you type in the name of the new folder.

To summarise, in order to be organised about working with computers, you need to:

- work systematically so that work can be found quickly
- save work in files with logical names
- create folders so that work of the same nature can be stored together
- produce back-up copies
- maintain good house-keeping, both in your work place and within your computer storage system.

Tidy up as you go and delete unwanted material so that you free up memory:

- Look after floppy disks and keep them away from magnets and extremes of temperature.
- Respect the equipment that you use and always follow correct procedures, whether it is for logging on or off or for reporting faults.

EXERCISE 11.8

This exercise is designed to show that you can use a computer in an organised way. It is **essential** that you are seen carrying out these tasks and following the instructions. Please also refer to Exercise 11.6.

You will need a short article from a newspaper or magazine to type out. You should **not** use the same article as other people in your group. The article should be about 250 words long.

Task 1
Switch on the computer, log-on (if you have to) and open a new file in a word processing format.

- Give a title to your article.
- Type it into the computer.
- **Ask** a supervisor to watch you use **spell-check**.
- **Change** both the font size and type.

Task 2
Using the same piece of typing:

- Go into Clipart (use *Insert* and follow instructions).
- Find a suitable picture.
- Add the image to your article at a convenient place. (You may need to make space.)

Task 3
When you have completed Tasks 1 and 2 you need to save your article.

- Save your article using a name that makes sense and will still make sense in several months' time.
- Put your new file in a new folder titled NVQ2IT.
- **Ask** a supervisor to witness that you are able to do this.

Task 4
This task requires you to manage files in your folder(s).

- Open a new file and type out some information. File it and then
- **Ask** your supervisor to watch you **delete** this file or

- Simply **delete a**n unwanted file but make sure you are seen doing this.
- Place a floppy disk in Drive A and then **transfer** your original file to the disk.
- Make sure that the disk has a label on it and that the label has an appropriate name.

Task 5
This task requires you to log-off following the accepted procedure for the computer that you are using.

- **Make sure** you have been witnessed carrying out all the tasks and that
- **Your supervisor** has signed the boxes in Exercise 10.5.

Task 6
Answer the following questions:

(a) My article was about

(b) I gave my file the following name:

(c) I have a back-up copy because

(d) It is important to delete unwanted files because

(e) Floppy disks should be stored away from

(f) Companies need to back-up their computer files because

(g) If I discovered that a computer was not working properly, the action I would take would be (explain how you report faults)

The above exercise has been completed satisfactorily by. . . . .

Signed . . . . .Job title . . . . Date . . . .

Accessing and using software packages

Engineering companies use a wide variety of software. Some software has general uses, for example, word processing software and spreadsheet programmes for financial management, while there is also a great deal of specialist software available. CAD and CNC software, in particular, may be used by engineering companies. For electronics, there are programs to design circuits, and many supply companies now issue their catalogues on CD-ROMs.

The types of software commonly used can be divided into overall groups such as **word processing, spreadsheets, databases, graphics, design** and many more. For your NVQ2 qualification, you need to show that you can use two of the following:

- Word processing
- Databases
- Spreadsheets
- Electronic communication (e-mail, intramail)
- Graphics.

You should be able to demonstrate some of what is required by providing a word-processed handout with graphics for the Communications–Short talk Assignment. However, you also need to know and understand quite a lot more about software packages and their use.

EXERCISE 11.9

Identify three different software packages used by your company.

Software	Used by (or for):
1.	
2.	
3.	

The above exercise has been completed satisfactorily by............................

Signed . . . . Job title . . . . Date . . . .

Types of software packages available

The major applications of the different types of software are listed below. We will then look at some issues in more detail.

A **word processing** package helps you to create documents. A word processor has largely replaced a typewriter and enables you

to do much more. Modern packages help you to produce very professional-looking documents. They come with on-screen HELP so that you can convert basic text into whatever format you choose. A great deal of the skill in using a computer is in having the confidence to try things out and make the most of the software.

A **spreadsheet** programme is mainly concerned with numbers and is most often used for business finances. The real usefulness of a spreadsheet package is that it allows you to manipulate the numbers you feed into the system and to do this very quickly. For example, you could easily set up one column containing the basic price of a component, the next column could have the amount of VAT payable and the next column the price plus VAT. You would only have to feed in the basic prices and enter a simple formula and the computer would do the rest. Spreadsheets will also present the information in the form of charts or graphs. Again, all that is needed here is for you to be able to follow on-screen Help.

Excel spreadsheet

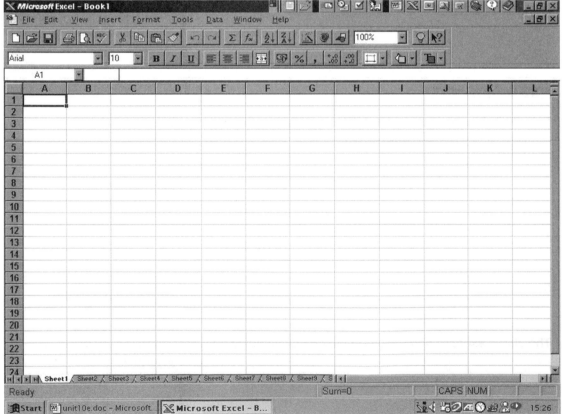

A database is a large collection of information. Like all information that is written down, it is most useful if it is organised in such a way that the information can be found easily and quickly. The simplest and easiest operation a database programme will carry out is to sort the collection of information into alphabetical order. For an address book, this would be ideal but the information may be a bit more complicated than simple addresses. For example, a company may have a database of all their suppliers, details of the goods that are supplied and how quickly they can supply the goods. A good database programme would be able to sort out the name of the supplier who can supply the required goods in the shortest time. When you use a database programme, you need to learn how to use the following commands – **sort**, **find** and answer a **query**.

Database

Would you access the supplier database please

Graphics packages are much used in engineering for a variety of purposes. These packages enable the user to use the computer to make drawings. As with the other types of software, you will need to practise to become skilful. Games packages include a lot of graphics and the more expensive the package, in general terms, the better the quality of the picture produced. Some graphics programmes are for very special applications. For example, you may use some software to help you design electronic circuits. In some cases, specialised software is first used to design a component and then the software interacts with a CNC machine to shape the component (CNC stands for Computer Numerical Control and the operation depends on the computer sending numerical coordinates to the lathe or mill so that it machines the component very accurately).

Electronic communication involves sending messages from one computer to another as electronic mail (e-mail) or as intramail in which the message is passed within a network rather than involving the World Wide Web and an Internet provider. One thing that you do need to remember here is that you have to be very accurate with the address to which you send information. An advantage of 'snail mail' or ordinary post is that even quite vague addresses will often work.

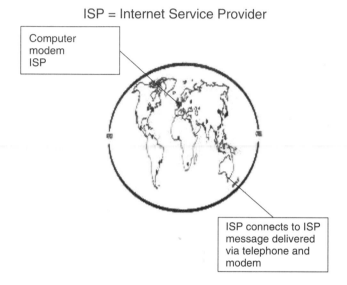

ISP = Internet Service Provider

Computer
modem
ISP

ISP connects to ISP
message delivered
via telephone and
modem

EXERCISE 11.10

You need to demonstrate that you can use computer software packages to aid engineering activities. Use a highlighter on the table below to identify the software applications you use and the general nature of the evidence you can produce. You need to be proficient with at least two of the following: word processing, spreadsheets, databases, graphics and electronic communication. Also, you need to show that you are capable of repeating the skill at least twice to the required standard.

Word processing	Spreadsheets	Databases	Graphics	Electronic communication
Letter	Adding up columns	Sorting data alphabetically	Adding graphics to a report	Sending basic message
Memo	Percentages	Using programme to find information	Creating a graphics image	Copying message to other people
Report plus a chart	Inputting a formula	Generating a query	Using colour	Adding attachments to a message

Report plus graphics	Using information for a purpose		Changing the size of an image	Filing a message that has been sent
Completed form	Generating a chart or graph		Adding text to an image	Deleting unwanted messages
Saving files	Saving files	Saving files	Saving files	
Deleting files	Deleting files	Deleting files	Deleting files	

The above exercise has been completed satisfactorily by.............................who has shown via questioning that s/he is able to provide the evidence required.

Signed . . . . Job title . . . . Date . . . .

We are going to concentrate on **word processing** and **electronic communication**. If you use a CAD package, such as Autosketch or AutoCAD, then you may provide evidence for **graphics**. You also need to be able to use the basic commands of both a spreadsheet package and a database package to meet key skills IT requirements.

Remember that before you can start using any software, you may need to log in and you will have to follow a set procedure. You may also have your own user name and password. If you have used a large computer system previously, you will know that there are often strict rules about computer use and abuse. A similar set of rules may apply where you work, and you are strongly advised to make sure you know and understand what these rules are. Abuse of computer facilities can lead to disciplinary procedures and should not be treated lightly.

Word processing program are designed to produce documents. Many program will interact within a computer so that you can transfer information from a graphics package (e.g. Clipart) or a spreadsheet into the document that you are developing.

Basic commands. A new 'page' for a word processing program is often called a document and needs to be named. The menu bar at the top of the screen holds most of the commands you need. Initially, you need to decide whether the writing should start at the left, in the middle or on the right.

Log-in name

Password********

Align left   Align middle   Align right

To choose, you 'click' on the icon for the format required. You may also want to choose the size and nature of the writing or **font**. Look at the example of the tool bar below. To change the size of the font, you click on the arrow to the right of the number 10 and a menu of different sized fonts will appear.

Menu bar

To change the font, click on **Format** and a menu of fonts will appear. Format can also be used to change font size.

Most documents are printed with the paper lengthways, but on occasions, you may want extra width and want to use the paper sideways.

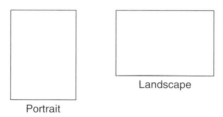

Landscape

Portrait

This is achieved by clicking on **File** and then choosing **Page Setup**. This menu gives you various options including Margins that sets the extent of clear space around the printed area, Paper size that allows you to select Portrait or Landscape, Paper Source and Layout.

Another useful icon is the **B**. When you click here, the print becomes **Bold**. The **I** converts the script to italics and the next icon. **U** underlines the text. Two more useful icons are the copy and paste icons. First, you use the mouse to highlight the text you want to copy and then you click on Copy. Next, you move the cursor (blinking vertical line) to where you want to place the copied text and then you click on Paste and the text appears in the new place. Because you have copied it, you may need to go back to the original position and delete the initial text.

To discover what the various icons mean, use the mouse to move the pointer under each icon in turn. From this you will discover **Bullets** (useful for summarising main points), **Numbering** (useful for identifying new points) and the various icons associated with sending documents to the printer, correcting spelling and saving the document. It is also possible to add icons to your toolbar. For an engineering student working with numbers, two useful icons are the **subscript x_n** and **superscript x^n**. Superscript is particularly useful for mathematics, for example, to write 10 cubed litres we simply type in 10^3 l.

Finally, the best way to get to know what your word processing programme is capable of is to practise and investigate. Click on icons and see what menus appear. Once you have mastered the basic commands, you are ready to produce a variety of documents.

Documents come in all shapes and sizes. Below are a selection of documents that have specific functions:

- Reports
- Formal letters
- Memos
- CVs
- Newsletters
- Proformas.

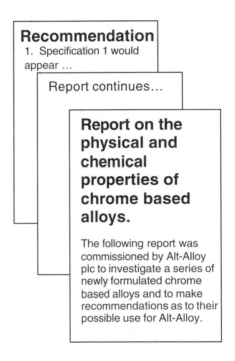

Reports start with a title, and then the actual report will depend on the nature of the information it contains. For example, most people remember writing up science experiments in school. These are a form of report that have definite sections such as prediction, apparatus used, method, results, conclusion. Reports need to be clear and organised. The end of a report may contain conclusions or recommendations. For example, a materials laboratory may have been testing different alloys prior to a component being manufactured. At the end of the report, we would expect there to be a recommendation as to which alloy best suits the application intended or a firm may have been researching the best type of CNC lathe to purchase; we would expect the last paragraph to make a recommendation. In the last example, we may expect the report to be set out with side headings as each lathe tested is commented upon in the report. Sometimes these sections are numbered.

Formal letters have a very definite format. The style of formal letter used by a company varies quite a lot but there is some essential information that must always be included. Look at the letter below and note this information:

DESIGNED FOR YOU

Unit 12
Priors Industrial Estate
WALLING
NV2 4QT

Date as postmark

Kwikzip Electronics
The Old Mill
OLDBURY
Hants
SS14 3HT

Dear Mr. Logan

Ref. Your quotation N/34/Jan 2000

We are pleased to accept your quotation for the components required. In your letter you
stated that batches can be dispatched within 3 working days of order and we will require
10 batches per week starting on April 10th. If our product is taken up as our
Marketing Department suggests it will be, then you can anticipate additional orders from
the end of May.

Please confirm that you can meet our requirements and our first delivery date.

Yours sincerely,

Andrew McCulloch
Production Manager

EXERCISE 11.11

Use a word processing program to write a formal letter from your firm's address to:
Mr S. Patel, Managing Director, New Forms Engineering, Unit 10, Aztec West Business Park,
Leeds, LS6 2PD.
Thank Mr Patel for the drawings and state that No. 10 seems ideal but that some changes
need to be made and that you will phone his secretary to make an arrangement for you to
visit him again to discuss these changes. End the letter in a formal way, leave six lines of
space for your actual signature and then add your name. Look at the letter above to make
sure you have your letter set out properly.

Letter to Mr Patel
State location of printout of letter, filename, path and back-up location for letter.

This confirms that. . . . . .has used a word processing package
satisfactorily to produce a formal style letter.

Signed . . . . .Job title . . . . Date . . . .

Memos (short for memoranda) are letters that are sent within a firm or organisation. To make life easier, they are often set out in a formal manner. Look at the example below in which Mr. Patel sends a memo to inform the design team that their work has been accepted but some adjustments are needed.

Memorandum

From: S Patel
To: Mark Water – Design Team
Date: 12th August, 2000

Re. Drawings for ABC Co. Ltd.

Our drawings have been accepted and they especially like No. 10. However, they do want some modifications and will contact us by phone for an appointment. Please ensure that you and Ez will be available even if you have to cancel something else.

Liz will contact you when they phone with a date.

EXERCISE 11.12

Write a memo that could be sent to your supervisor telling him/her that you have been asked to visit another company by the Training Manager to watch a health and safety demonstration and that you will be away from work for an afternoon. (Make sure you give the actual date involved.)

State location of printouts of memo, file name, path and back-up location for this memo.

This confirms that. . . . . .has word processed the memo above to a satisfactory standard.

Signed . . . . Job title . . . . Date . . . .

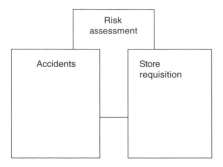

A proforma is a document that needs to be filled in with specific information. In an earlier chapter, you were asked to complete an accident proforma. These documents are useful because they ensure that all the required information is provided. They can be very detailed or as simple as a requisition slip for some material from the stores.

EXERCISE 11.13

Identify three forms that you know are used in your company.

Three forms used where I work are:

1.

2.

3.

Two forms I use on a regular basis are:

1.

2.

The above exercise has been completed satisfactorily by............................

Signed . . . . Job title . . . . Date . . . .

Getting the most out of the computer to produce documents

As you have seen, there are different types of documents and you can make use of the computer to produce a document in the format you need.

Using the tool bar and scroll down menus. One of the most useful things you should learn to do when using a computer is to make use of the on-screen information. At the top of the screen you will find the **tool bar**:

Menu bar

To use the tool bar you need to use the mouse to left click on one of the prompts, for example, File. A menu appears below the tool bar prompt, and what you should have learnt from working through the section above is that with little effort you can get a lot out of your computer. You need to learn to use the icons and the scroll down menus and follow the on-screen prompts. If you go wrong, click on Edit and the first choice it gives you is to undo what you have just done.

Once you are satisfied with your work on screen, you can send it to the printer. To do this, you have two choices – click on the Printer icon along the tool bar or click on File and choose the Print option. If you want to print more than one copy, it is usually easier to click on the Print option as you then get an option as to how many copies you want to print.

Print

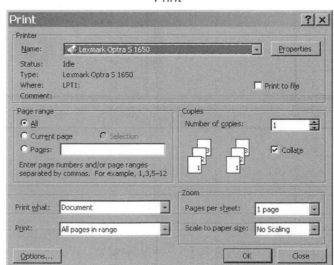

If you look carefully at the screen above, you can see that you have a lot of options. Under **Page range**, you can choose to print either all the pages, the current page, which is the page in front of you on the screen, or selected pages. In the Copies section, you can choose the number of copies you want to make and there are other options as well. By making the best use of this help screen, you can save paper by printing copies back-to-back. To do this, you choose current page, send this to the printer and then pull up the next page on the screen in front of you. Once the printer has finished printing the first page, you leave the ink time to dry and then put the same page back in the printer so that the second page can be printed on the reverse. The worst that can happen is that you print on the same side of the paper but a little practice and attention to the printer will quickly sort out what you should do!

Finally, having taken a lot of effort to make an excellent document, you need to ensure that it is saved safely and in a way that it can be found again.

Occasionally, you may want to stop other people accessing important documents. We have looked at personal codes but it is also possible to give extra protection to a document using the **Tools** icon.

Protect document

Protect Document [?] [X]

Protect document for

(•) Tracked changes

( ) Comments

( ) Forms: [Sections...]

Password (optional):

[   ]

[OK]   [Cancel]

You can make sure that no one changes the document without it being registered and also you can have the added protection of an extra password.

Electronic communication

One of the most exciting developments in electronic engineering and communications engineering has been the development of the Internet and the World Wide Web, and the story is far from over. The latest developments involve mobile phones and their access to the World Wide Web and we shall see a great deal of increasingly sophisticated technology in this sphere before very long.

Without going into too much detail, it is worth taking a quick look at the web and how it works. The idea for the web came from the United States Defence Department in the 1960s during the Cold War. The idea was to link computers holding sensitive information by telephone in a net-type arrangement so that if one computer was destroyed, information could still be passed easily by rerouting via another path in the web. Many people initially involved in the web were scientists and they quickly appreciated that information other than defence material could be passed quickly to other scientists using this new network. The main problem at this time was that computers were in the early stages of development and only very large organisations owned them. Since then, and particularly within the last ten years, computer technology has developed at such a pace with the result that they have become affordable for individuals.

So, how does the Internet work?

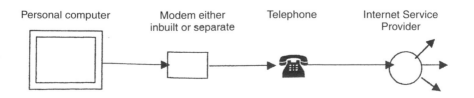

Personal computer   Modem either   Telephone   Internet Service
                     inbuilt or separate   Provider

As you can see from the diagram, all you need is a personal computer, the correct software, a modem and a telephone line. To access the Internet, you need to be connected to an Internet Service Provider (ISP). Any ISP has specialist server software that allows the people connected to it to browse through the services it has to offer and make other connections. Some connections will be via telephone lines but, increasingly, communications technology uses fibre optic cables.

To access a web site, you need the web site's address. All web site addresses start with http://, which stands for hyper text transfer protocol. The http://, is followed by www. or world wide web and this is followed by more information. The whole address can be referred to as its *URL or Uniform Resource Locator*.

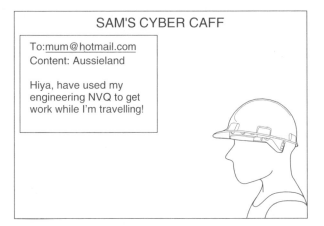

Many people like the Internet simply for e-mailing their friends or to find out about services that are offered via the internet. E-mails are part of the Internet and use slightly different software from the rest of the web. Many firms now have e-mail addresses for all their staff and people use e-mails as often as they use the telephone.

Web addresses referred to as **dot coms** have hit the headlines, so what does this mean? Basically, you can work out quite a lot about the web address by looking at its various parts. If you look at the back of this book you will see the address *http://newnespress.com* The book is published by Butterworth-Heinemann, which is part of Newnes Press and they are a commercial publisher hence the **.com**. If you study at a college or have friends at university or school and send them e-mails, you will use **.ac** as part of the address. The **.ac** indicates an academic institution. The following also occur very frequently, **.org** and **.co** and **.gov**. This last one indicates that it is part of the government. These are referred to as *domain names*.

You can also tell which country the web address is in as **.uk** means United Kingdom, **.fr** indicates France and **.au** is Australia.

You may wonder how it is all paid for. When you connect to the web, you are using a telephone connection that you are paying for. Your ISP probably receives a percentage of this call fee to provide the

Internet service. Free services to you, once you are connected to the web, are paid for by the advertisers on the web. There are also some parts of the web that you can only access if you pay a subscription.

Anyone with a computer and the correct software can have a web address and you can put any information into your web site that you choose. Increasingly, a large number of small firms and organisations are setting up web sites to advertise their products and services and of course they include e-mail addresses so that people who have browsed their pages can contact them to place orders or request additional information.

Internet accuracy

One thing that you should bear in mind is that there is no check on the quality or accuracy of the information on the web. Do not think that there is anything really special about information taken from the web. It is only as good as the person who wrote it. The main advantage of good web information is that it can be very up-to-date. For example, satellite photographs taken today can be on the web the same day, whereas it takes a long time to get a book into print.

Other facilities available using the web include newsgroups, chat rooms, shopping and entertainment. If you have the right software and a sufficiently powerful computer (otherwise it takes too long) you can also listen to music, watch video clips and work through interactive sites.

Mobile phone manufacturers are promoting access to the web using a mobile phone. These phones use **wap** or wireless application protocol. Because the screens on mobile phones are small, the **wap** pages are reduced. (If you look back to the beginning of this unit, you can remind yourself about reducing information.)

Using the internet.   You can use the Internet as a private individual through your personal computer or through work. Whichever is the case, you will need to log on, which will take you to the first web page of your ISP. These pages are interactive in that you use the mouse to click on on-screen prompts.

Global internet home page

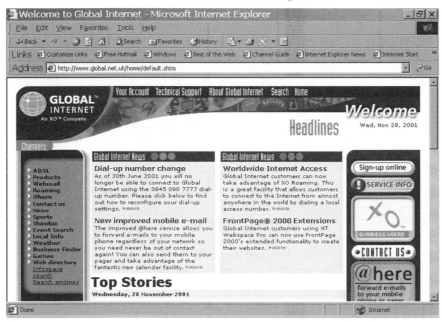

The best way to find out how to get the best out of the Internet is to 'have a go'. Alternatively, you can have a guided tour by someone who is already experienced at using the Internet or you can take a course at a local college or library.

Sending and receiving e-mails – ISPs provide an e-mail service. There are now so many ISPs with so many offers that it is impossible to give one good example. The important things to remember about e-mails are that you will need:

- a **log-in name**
- a **password**.

Once you are logged in, the usual procedure is to check to see if you have new mail. Most systems show you a screen with the mail you have received. To read the full e-mail, you need to click on the letter you want to open and then the first screen disappears and your 'letter' is in front of you. When you have read it, you can do a number of things that are outlined on a tool bar. You can:

- **Reply** – To do this you usually need to click on **Reply** and a new screen will appear. The address will already be in place and you will need to go to the next box in which you write down the subject of your e-mail. Under this box, there will be another box with **cc** in it. This allows you to copy the letter to other people.
- **Delete** – This gets rid of the letter. You do need to manage your e-mails, otherwise you may get too many to cope with.
- **Forward** – Here the e-mail is forwarded to another address.
- **Move to folder** – Just as you have learnt how to open folders to keep your other computer files tidy, you can have e-mail folders in which you can keep letters from different people or on different topics in separate files.
- **Compose** – One of the problems of using the **reply** facility is that not only does your reply but also the original letter that you received

go to the person. If you choose **compose**, you can still reply to a letter without the original going back as well.

- **Attach** – You may want to send other information that is stored elsewhere in your computer. When you click on the **attach** icon, you will receive on-screen instructions to help you find the file you want to attach. Sending attachments allows you to prepare information in advance, while not on-line.

Working off-line – All the time you are on the Internet, you are using a phone line and this is costing you or your employer money. A good way to work is to work off-line and then attach to the Internet when you are ready. Complicated files and engineering drawings can be prepared in advance and then sent as an attachment to a much shorter e-mail.

Free e-mail web sites – There are some web sites that offer free e-mail facilities. These have the advantage of being capable of being accessed anywhere in the world from any computer and you do not need to go through your normal ISP if you are travelling. What it does mean is that you will have more than one e-mail address and you will have to remember another log-in name and password.

EXERCISE 11.14 Electronic communication checklist

If you have chosen to demonstrate that you are proficient at **electronic communication** you now need to demonstrate that you are able to do the following and you should be watched carrying out the following at least twice:

Criteria:	Witnessed by:	Date:
Accessing e-mail facilities		
Sending basic message		
Copying messages to other people		
Adding attachments to a message		
Filing a message that has been sent		
Deleting unwanted messages		

The above exercise has been completed satisfactorily by...........................

Signed................Job title.................................Date..........

Viruses are programs that upset the normal running of a computer. Viruses take a number of different forms. Serious viruses can destroy information on the hard drive. A common way of spreading viruses is via attachments to an e-mail. You cannot import a virus by reading an e-mail but if you open an attachment, you must be aware that you can import a virus. Also, if you file an e-mail elsewhere in your system, you can import viruses. Viruses are a scourge that can cost the industry an immense amount of money. There is antivirus software available and all computer users are advised to make use of it.

Spam or junk mail – If you e-mail commercial sites, then your e-mail address becomes known and you might receive junk mail. Many ISPs provide services that help you to manage junk mail. To do this, look for on-screen tips. Junk mail is usually the same thing that you receive through the post except that this time it finds you electronically.

Surfing the Internet or using the Internet to provide information is relatively easy but as you become more familiar with search skills, you find the information you require faster.

There are three main ways of accessing information:

- By typing in the web address, for example, http://www etc. and then clicking on 'Go'.
- By typing in a topic in a query box. Most ISPs have a preferred Search Engine, for example, UK Gold.
- By typing in the web address such as 'Yahoo' or 'Ask Jeeves' and then following on-screen prompts.

EXERCISE 11.15

Use the Internet. Try out the three ways listed above.
(As web sites change, the following are suggestions that can be adapted.)

Type of query and instructions	What to find out	Nature of information, e.g. *see printed material on*, or *what I found out about. . . was*
1. Type in *www.rolls-royce.com* and then find out about 'Careers'.	Find out about where in the world Rolls-Royce employs people or some alternative information.	
2. Use the ISP query box to find out about the London Eye.	Some of the firms involved in building the 'Eye'.	
3. Type in *www.yahoo.com*. When the yahoo site comes up, find 'Engineering' and then 'Automotive'.	Find out about concept cars and alternative fuel vehicles or state other query.	

This confirms that. . . . . .has demonstrated satisfactorily that s/he is able to use basic search skills on the Internet.

Signed . . . . .Job title . . . . Date . . . .

Using spreadsheets

Spreadsheet programmes were originally developed for business purposes and their main use is still to help companies manage their finances. However, spreadsheet programs basically carry out arithmetic and can be used for a variety of purposes as well as financial management.

There are a number of spreadsheet programs but a common one is Microsoft Excel.

Excel spreadsheet Book 1

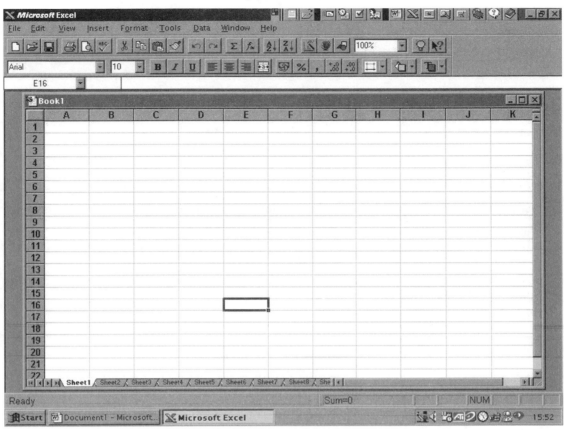

A fresh 'page' for a spreadsheet is called a *book* and if you look at the diagram above, below the tool bar you will see Book 1. The rectangle in the diagram is one **cell** and there are **rows** and **columns** of cells. Each cell can be located from a grid reference. The highlighted cell is E16, that is, in **Column E** and **Row 16**.

An **active cell** is one that is bordered like E16 and this is the cell to which information is being added or can be added. The **arrow** keys on the keyboard are used to move from cell to cell or you can use the mouse and 'click' on the cell you want to become active.

In the spreadsheet shown below, you can see that 'Designed for You' make some weird products and some of the details are shown. The information on the left (that is not shaded) is basic information that needs to be fed in. The information that is shaded was all calculated by the computer by using formulae. To find the total cost of an item, materials and labour need to be added together. In this case, the sum is very easy, but to do it on the computer what you do is:

- click on cell E7 to make the cell active
- go to the tool bar and click on Σ(= autosum)
- at the top of the page, above Book 1, E7 will appear and next to it Sum(B8)
- in this case, what you want to know is the total price, which is cell C7 added to D7.

Excel spreadsheet with production costs

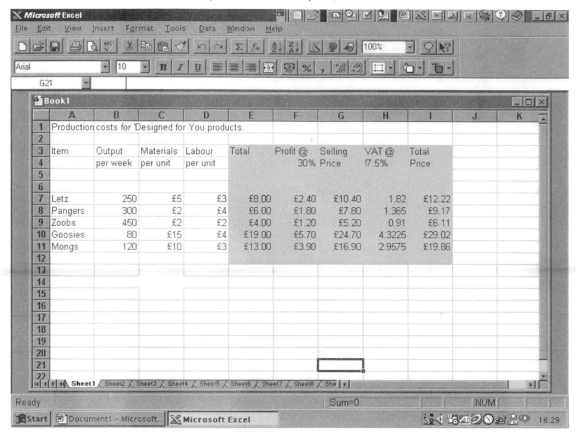

Databases for engineering

A database is a large collection of information and a database program helps you to manage this information. As we saw with file management, it is important to store information logically and efficiently so that it can be found again very quickly. Look at the following information about a customer for 'Designed for You':

First name	Last name	Address 1	Address 2	Post code	Last order	Tel No.
James	Page	4 Link Rd	Cardiff	CF32 9HJ	01.05.00	0122499763

If we had a paper-based filing system, we may ask ourselves which of the bits of information are most important to give a clue where to file it. We may decide to do quite a lot of photocopying. With a database computer programme, we do not need to do this.

Managing the information. **Sort** is the command used to organise the information. For example, you could sort all employees alphabetically by their last name. In this case, you would also have to tell the programme to include the first names as well. The instruction you would give would be:

Sort on LAST NAME and FIRST NAME.

Find is the command used to locate the information you need. For example, often if you make a telephone enquiry about something, you will be asked your name and postcode and the program user will then **find** your address.

You can also ask the program questions. This is called a *query*. The sales department may want to target customers who have bought a certain product over the last five years with information regarding a new version of the product. The program would effectively sieve through all the information and identify and display those customers required for the mail shot.

The way you formulate the query depends on the program you are using.

What a database looks like

			Table: a collection of information about a specific topic			A field name identifies the nature of the information in a field

First name	Last name	Address 1	Address 2	Postcode	Tel. No
Sean	McCready	1 Green Lane	Romford	SS67 2RF	0134632765
Natasha	O'Leery	3, Tile Hill	Coventry	CO12 8HB	0178486395
Liam	Brown	5, Nut Street	Bradford	BF23 6VC	0176399765
Robert	Smith	7 Hill Head	Burnley	BB20 8PL	0189643298
Susan	Smith	9 Grove Road	Burnley	BB12 5FK	0189656332
Rajah	Patel	11 Proud Street	Leeds	LS8 4FD	0199877654
David	Carp	13 Bell Lane	Blackburn	BB10 9DR	0187465332
Ali	Iqbar	15 Kite Place	Preston	PR22 8AD	0167543388
Flora	McDonald	17 Palace Road	Carlisle	CL2 4LM	0122388765

A field is one piece of information and comprises part of a record

A record is a collection of information about one topic, e.g. details of the company

Entering Information into a database depends on the program you are using. If a new record needs to be added to the database above by following on-screen instructions for adding to addresses, a **form** will appear.

ADDRESSES	☐ ☐ ☒
First name	
Last name	
Address 1	
Address 2	
Postcode	
Telephone Number	

Using databases. There are different types of databases and the type set up often depends on the use to which it is being put. For example, most firms will have an internal telephone extension list that needs to be updated regularly. A **flat file database** will be set up for such a purpose as it can easily be updated. It may look like this:

EMPLOYEE TELEPHONE NUMBERS

NAME	DEPARTMENT	EXTENSION NO.
ADAMS, Alan	Accounts	234
BERNARD, Cindy	Personnel	542
CROW, Matt	Maintenance	221
DAVIES, Karen	Stores	563
EVANS, Dewi	Maintenance	227

A **relational database** combines information from separate sources. For example, a firm may have received a list of addresses of customers from a sister company with the group (i.e. it receives a **table**). This table could then be combined with an existing table so that potential customers could be identified. Such a database is much more difficult to set up.

A common use of a database for an engineering company would be a database held by the stores department. Nowadays, firms try not to hold too much stock in their stores as stock has to be paid for and while items are on shelves they represent money that is not doing anything. Just-in-time ordering ensures that items are available when they are required but are not in the stores for any length of time. A database of items is set up and can be organised to generate orders when stock levels reach a critical point. When someone goes to the stores, the store person can check from the database whether something is available and how many are still in stock. The location of items can also be held in the database so that locating the item on the shelves is easy.

EXERCISE 11.16   Using databases

Identify two databases that are used in your company:	1. 2.

If you added extra information on a new client to a database you would be adding:

A an extra record
B a table
C a new column
D a new field

If you wanted to use the database to sift out customers who had all received the same product over the last year, you would use:

A find
B sort
C query
D all of the above

The above exercise has been completed satisfactorily by. . . . .

Signed . . . . Job title . . . . Date . . . .

12 Problem solving

Exercise checklist

(Ask your assessor to initial the lower box for each exercise when you have completed it.)

Exercise no.	12.1	12.2	12.3	12.4
Initials				
Date				

In your work as an engineer, you normally carry out numerous tasks using different equipments and make decisions as to what tools and materials to use. You also follow instructions, but a job is never quite that simple. In Chapter 3, you learned that you sometimes have to make decisions when information is missing. This is using your initiative and when you do this, you are solving problems as well.

It is often said that engineers are professional problem solvers.

A typical workshop problem might arise when machining unfamiliar materials or using non-standard tooling. Problems can arise in a wide variety of situations and an even larger array of possible solutions can be initiated to overcome them.

Consider the Case study below:

Case study

DS & PP Engineering Ltd have begun work on grinding the profile of some turbine blades that are made of a special nickel alloy called *nimonic*. This material is tough, heat resistant and does not stretch. When grinding the nimonic material, the operator found that the required surface finish was difficult to achieve. The matter was discussed with the foreman but no easy solution was found.

Problem title: Surface finish on nimonic

There are two possible ways of finding a solution to this problem:

- Adjust the feeds and speeds of the grinding operation.
  This action was carried out, but the surface finish was still rougher than that allowed on the drawing and the problem persisted. It then became necessary to initiate a second action.
- Contact experts for advice.

In this case, both grinding wheel manufacturers and cutting fluid suppliers can be consulted for process advice.

Representatives from the grinding wheel and the cutting fluid companies would need to know the exact problem and may arrange to visit the works and observe the process. You will need to give them as much information as possible about the task to enable them to make any recommendations.

Information gathered would include the following:

Materials	*Process*	*Requirements*	*Grinding wheel*
Work material	The machine used, its age, type and condition	Surface finish required	Diameter and rev/min
Cutting fluids used	Workpiece feed rates	Workpiece size	Grit size, type and bond
	Evaluating results gathered	Workpiece shape	

After the visits, the recommendations are reviewed and some selected for trial. The trials are then carried out, costed and the results evaluated against the work specification. If the outcome is satisfactory, the problem is confirmed as having been solved and all process sheets would be amended.

From this case study, you can see that there are three main stages in problem solving:

1. Identifying the problem accurately and exploring possible ways of solving it
2. Trying out one or more possible solutions
3. Evaluating the results to check that the problem has been solved.

Note that we always learn from the experience of solving a problem.

Because so much of engineering is solving problems, you may not realise how many problems you solve while doing your normal duties. In the chapters on fitting and machining (Turning and Milling) you would have come across job sheets and other ways of doing a task if the recommended way is not possible. These were examples of you solving problems. Apart from being able to solve problems, you need to demonstrate this skill.

The following forms have been set out to assist you in generating evidence towards key skills 'problem solving'.

EXERCISE 12.1 Identification of a problem

Identify at least **two** problems encountered in your engineering activities. The categories here are suggestions and by no means exhaustive.

- Tools not available
- PC software problems
- Deficiencies in information
- Working with others
- Equipment failure
- Communication 'mix ups'

I have encountered problems with the following:		
1.	2.	3.

Discuss your problems with your supervisor and ask him/her to validate them as sufficiently complex for the level required by signing the witness testimony form below.

Witness testimony

I confirm that the problems outlined above were encountered by
. . . . and that they represent a complexity consistent with key skills: problem solving at Level 2.

Signed . . . . Job title . . . . Date . . . .

EXERCISE 12.2

Now that your problems have been given titles, list the critical features of at least two problems.

Title 1	Title 2
Features of problem: (a) (b) (c)	Features of problem: (a) (b) (c)

EXERCISE 12.3

Produce two alternative method routes of overcoming each problem (you may use the work planning sheet on Appendix XVI for this).

Title:		Title:	
Route A	Route B	Route A	Route B

State which route is the preferred solution from your options and give clear reasons for your choices. If you asked for help and/or advice on solving your problem, indicate from whom the advice was given.

Title:	Title:
Route selected	Route selected
Reason for selection	Reason for selection
Help/advice was received from:	

Ask your supervisor to approve your plan. Enter into the space any additions/alterations/modifications that were recommended.

Witness testimony

I confirm that the plans above devised by . . . . have been discussed with me and represent a suitable method of overcoming the problem.

Title 1:	Title 2:
Additions/alterations/modifications recommended:	

Signed . . . . .Job title . . . . Date . . . .

EXERCISE 12.4   Reviewing problems solved

Was the job affected in any way? For example:

- Final cost
- Time delay
- Accuracy
- Location
- Additional resources required
- Other factors.

Title 1	Title 2
Factors affecting my job	

Ask your supervisor to confirm that the solution applied to your problems enabled the problem to be solved. You need to prove that the problems have been solved; this can be done by showing the supervisor evidence. Your evidence may be:

- revised route card
- inspection sheet
- reference to the critical features of the job.

Witness testimony

I confirm that the problems outlined above were solved by . . . . ,
and that the evidence shown to me indicates that the solution was effective and efficient.

Signed . . . . Job title . . . . Date . . . .

Appendix I

Millimetres to inch conversions

mm	Inch	mm	Inch	mm	Inch	mm	Inch	mm	Inch
0.01	0.0004	7	0.2756	31	1.2205	55	2.1654	79	3.1102
0.02	0.0008	8	0.3150	32	1.2598	56	2.2047	80	3.1496
0.03	0.0012	9	0.3543	33	1.2992	57	2.2441	81	3.1890
0.04	0.0016	10	0.3937	34	1.3386	58	2.2835	82	3.2283
0.05	0.0020	11	0.4331	35	1.3780	59	2.3228	83	3.2677
0.06	0.0024	12	0.4724	36	1.4173	60	2.3622	84	3.3071
0.07	0.0028	13	0.5118	37	1.4567	61	2.4016	85	3.3465
0.08	0.0032	14	0.5512	38	1.4961	62	2.4409	86	3.3858
0.09	0.0035	15	0.5906	39	1.5354	63	2.4803	87	3.4252
0.1	0.0039	16	0.6299	40	1.5748	64	2.5197	88	3.4646
0.2	0.0079	17	0.6693	41	1.6142	65	2.5591	89	3.5039
0.3	0.0118	18	0.7087	42	1.6535	66	2.5984	90	3.5433
0.4	0.0158	19	0.7480	43	1.6929	67	2.6378	91	3.5827
0.5	0.0197	20	0.7874	44	1.7323	68	2.6772	92	3.6220
0.6	0.0236	21	0.8268	45	1.7717	69	2.7165	93	3.6614
0.7	0.0276	22	0.8661	46	1.8110	70	2.7559	94	3.7008
0.8	0.0315	23	0.9055	47	1.8504	71	2.7953	95	3.7402
0.9	0.0354	24	0.9559	48	1.8898	72	2.8346	96	3.7795
1	0.0394	25	0.9843	49	1.9291	73	2.8740	97	3.8189
2	0.0787	26	1.0236	50	1.9685	74	2.9134	98	3.8583
3	0.1181	27	1.0630	51	2.0079	75	2.9528	99	3.8976
4	0.1575	28	1.1024	52	2.0472	76	2.9921	100	3.9370
5	0.1969	29	1.1417	53	2.0866	77	3.0315	250	9.8430
6	0.2362	30	1.1811	54	2.1260	78	3.0709	1000	39.370

Appendix II — Inch to millimetres conversions

Fractional inch	Decimal inch	mm	Fractional inch	Decimal inch	mm	Fractional inch	Decimal inch	mm
–	0.001	0.0254	–	0.2	5.080	$\frac{5}{8}$	0.6250	15.88
–	0.002	0.0508	$\frac{13}{64}$	0.2031	5.16	$\frac{41}{64}$	0.6406	16.27
–	0.003	0.0762	$\frac{7}{32}$	0.2188	5.56	$\frac{21}{32}$	0.6562	16.67
–	0.004	0.1016	$\frac{15}{64}$	0.2344	5.95	$\frac{43}{64}$	0.6719	17.07
–	0.005	0.1270	$\frac{1}{4}$	0.2500	6.35	$\frac{11}{16}$	0.6875	17.46
–	0.006	0.1524	$\frac{17}{64}$	0.2656	6.75	–	0.7	17.780
–	0.007	0.1778	$\frac{9}{32}$	0.2813	7.14	$\frac{45}{64}$	0.7031	17.86
–	0.008	0.2032	$\frac{19}{64}$	0.2969	7.54	$\frac{23}{32}$	0.7188	18.26
–	0.009	0.2286	–	0.3	7.620	$\frac{47}{64}$	0.7344	18.65
–	0.01	0.254	$\frac{5}{16}$	0.3125	7.94	$\frac{3}{4}$	0.7500	19.05
$\frac{1}{64}$	0.0156	0.39	$\frac{21}{64}$	0.3281	8.33	$\frac{49}{64}$	0.7656	19.45
–	0.02	0.508	$\frac{11}{32}$	0.3438	8.73	$\frac{25}{32}$	0.7813	19.84
–	0.03	0.762	$\frac{23}{64}$	0.3594	9.13	$\frac{51}{64}$	0.7969	20.24
$\frac{1}{32}$	0.0312	0.79	$\frac{3}{8}$	0.3750	9.53	–	0.8	20.320
–	0.04	1.016	$\frac{25}{64}$	0.3906	9.92	$\frac{13}{16}$	0.8125	20.64
$\frac{3}{64}$	0.0469	1.19	–	0.4	10.160	$\frac{53}{64}$	0.8281	21.03

Fractional inch	Decimal inch	mm	Fractional inch	Decimal inch	mm	Fractional inch	Decimal inch	mm
–	0.05	1.270	$\frac{13}{32}$	0.4063	10.32	$\frac{27}{32}$	0.8438	21.43
–	0.06	1.524	$\frac{27}{64}$	0.4219	10.72	$\frac{55}{64}$	0.8594	21.83
$\frac{1}{16}$	0.0625	1.59	$\frac{7}{16}$	0.4375	11.11	$\frac{7}{8}$	0.8750	22.23
–	0.07	1.778	$\frac{29}{64}$	0.4531	11.51	$\frac{57}{64}$	0.8906	22.62
$\frac{5}{64}$	0.0781	1.98	$\frac{15}{32}$	0.4688	11.90	–	0.9	22.860
–	0.08	2.032	$\frac{31}{64}$	0.4844	12.30	$\frac{29}{32}$	0.9063	23.02
–	0.09	2.286	$\frac{1}{2}$	0.5000	12.70	$\frac{59}{64}$	0.9219	24.42
$\frac{3}{32}$	0.0938	2.38	$\frac{33}{64}$	0.5156	13.10	$\frac{15}{16}$	0.9375	23.81
–	0.1	2.540	$\frac{17}{32}$	0.5312	13.49	$\frac{61}{64}$	0.9531	24.21
$\frac{7}{64}$	0.1093	2.78	$\frac{35}{64}$	0.5469	13.89	$\frac{31}{32}$	0.9688	24.61
$\frac{1}{8}$	0.1250	3.18	$\frac{9}{16}$	0.5625	14.29	$\frac{63}{64}$	0.9844	25.00
$\frac{9}{64}$	0.1406	3.57	$\frac{37}{64}$	0.5781	14.68	1	1.0000	25.40
$\frac{5}{32}$	0.1563	3.97	$\frac{19}{32}$	0.5938	15.08	2	2.0000	50.80
$\frac{11}{64}$	0.1719	4.37	–	0.6	15.24	5	5.0000	127.0
$\frac{3}{16}$	0.1875	4.76	$\frac{39}{64}$	0.6094	15.48	10	10.000	254.0

Tapping sizes for common threads

The tapping sizes in this chart are for standard threads only.

ISO Metric (Thread angle 60°)			BSW (Whitworth) (Thread angle 55°)			BSF (British Standard Fine) (Thread angle 55°)		
Major diameter	Pitch mm	Tapping size	Major diameter	Pitch tpi	Tapping size (mm)	Major diameter	Pitch tpi	Tapping size (mm)
M4 × 0.7	3.3		1/8	40	2.55	–	–	–
M5 × 0.8	4.2		3/16	24	3.7	3/16	32	4.0
M6 × 1.0	5.0		1/4	20	5.1	1/4	26	5.3
M7 × 1.0	6.0		5/16	18	6.5	5/16	22	6.8
M8 × 1.25	6.8		3/8	16	7.9	3/8	20	8.3
M10 × 1.5	8.5		7/16	14	9.3	7/16	18	9.7
M12 × 1.75	10.2		1/2	12	10.5	1/2	16	11.1
M14 × 2	12.0		9/16	12	12.0	9/16	16	14.0
M16 × 2	14.0		5/8	11	13.5	5/8	14	12.6

BA (British Association) (Thread angle $47\frac{1}{2}°$)			UNC (Unified Coarse) (Thread angle 55°)			UNC (Unified Fine) (Thread angle 55°)		
Designating No.	*Pitch tpi*	*Tapping size* (mm)	*Major dia.*	*Pitch tpi*	*Tapping size* (mm)	*Major dia.*	*Pitch tpi*	*Tapping size* (mm)
0	25.4	5.0	1/4	20	5.2	1/4	28	5.4
1	28.2	4.4	5/16	18	6.5	5/16	24	6.8
2	31.3	4.0	3/8	16	8	3/8	24	8.4
3	34.8	3.3	7/16	14	9.3	7/16	20	9.8
4	38.5	2.95	1/2	13	10.8	1/2	20	11.4
5	43.1	2.6	9/16	12	12.1	9/16	18	12.8
6	47.9	2.25	5/8	11	13.5	5/8	18	14.5

The sizes of the tapping drills listed above are the nearest convenient size larger than the thread's core diameters. If you need to use an alternative drill, use the next larger size.

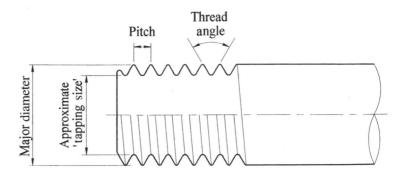

Appendix IV

Length of chords for various pitch circle diameters

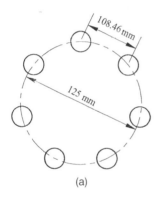

(a)

No. of holes	Pitch of holes	No. of holes	Pitch of holes
3	0.8660	17	0.2588
4	0.7071	18	0.2393
5	0.5878	19	0.2225
6	0.5000	20	0.2079
7	0.4339	21	0.1951
8	0.3827	22	0.1837
9	0.3420	23	0.1736
10	0.3090	24	0.1646
11	0.2817	25	0.1564
12	0.2588	30	0.1490
13	0.2393	35	0.1423
14	0.2225	40	0.1362
15	0.2079	45	0.1305
16	0.1951	50	0.1253

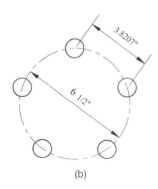

(b)

To convert the above values into chord lengths for circles with a diameter other than one unit (i.e. 1 mm or 1″), multiply the pitch circle diameter (PCD) by the number extracted from the table.

EXAMPLE 1
To find the chord length for marking out seven holes on a 125-mm PCD.

$0.8677 \times 125 = \mathbf{108.4625\,mm}$

This figure can be approximated to 108.5 mm.

EXAMPLE 2
To find the chord length for marking out five holes on a $6\frac{1}{2}''$ PCD.

$0.5878 \times 6\frac{1}{2} = \mathbf{3.8207''}$

This figure can be approximated to 3.82″ or 3 13/16″.

Appendix V

Example orthographic drawing (first-angle projection)

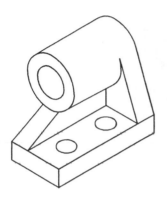

The isometric view of the pulley mount has been redrawn in first-angle orthographic projection in the frame at the bottom of the page. Imperial dimensions are used, a sectional view is added and correct types of lines are shown.

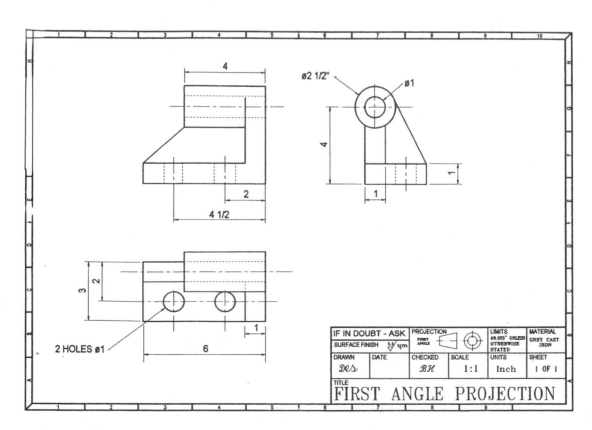

Example orthographic drawing (third-angle projection)

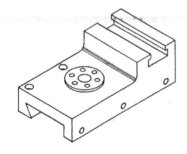

The isometric view of the slideway has been redrawn in third-angle orthographic projection in the frame at the bottom of the page. Metric dimensions are used and correct types of lines are shown.

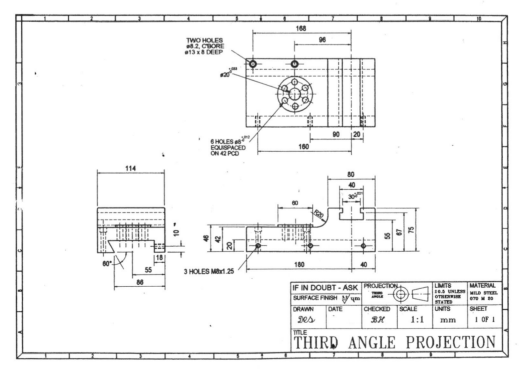

Drilling speeds and feeds

1. Drilling machine's spindle speed:

Drill diameter (mm)	Aluminium	Brass	Cast iron	Bronze	Mild steel	High-carbon steel	Hard alloy steel
Up to 3	5250	4800	850	3000	2500	2000	950
4–8	1600	2400	450	1500	1250	1000	475
8–10	2000	2000	310	1200	950	650	325
10–14	1400	1250	220	800	650	475	250
14–20	950	880	150	550	450	350	175
20–30	750	640	100	400	325	225	120
30–50	500	400	60	250	200	150	75

The above spindle speeds should only be set if your cutting conditions are ideal, that is, if you have sharp tools, cutting fluid, secure clamping and a rigid machine tool.

2. Drill feed rates:

Drill diameter (mm)	Feed rate (mm/rev)	Feed in/rev
Up to 3	0.05	0.002
4–8	0.06	0.003
8–10	0.15	0.006
10–14	0.20	0.008
14–20	0.30	0.012
20–30	0.40	0.016
30–50	0.50	0.020

The above feed rates should only be set if your cutting conditions are ideal (i.e. if you have sharp tools, cutting fluid available, secure clamping and a rigid machine tool). If your conditions are not ideal, use a slower speed.

Example:
When setting up to drill a ø12 hole in cast iron, the spindle speed should be set at 220 rev/min and the feed rate should be set at 0.2 mm/rev (0.008 in/rev).

Appendix VIII

Spindle speeds for turning and milling

To calculate the correct spindle speed for a machining operation, it is necessary to make a calculation using the formula shown below. The calculated revolutions per minute should be accepted **only** as nominal. Adjustments to this calculated speed can be made to determine the optimum speed during the component manufacture. Remember that if there is no cutting fluid or coolant available, the speed should be reduced.

Material	Cutting speed
Aluminium	60 m/min
Brass	45 m/min
Cast iron	25 m/min
Bronze	30 m/min
Mild steel	25 m/min
High-carbon steel	18 m/min
Hard alloy steel	10 m/min

Note. When using tungsten tools multiply the above values by 2.

$$\text{Rev/Min} = \frac{1000 \times S}{\pi \times d}$$

Where S = Cutting speed of the work material

d = Work (or tool) diameter

π = 3.142

This table shows the recommended cutting speeds (values for S) in metres per minute for some common engineering materials.

The above cutting speeds should only be used if your cutting conditions are ideal. (i.e. if you have sharp tools, cutting fluid available, secure clamping and a rigid machine tool). If your conditions are not ideal, use a slower speed.

Example 1: To **turn** a mild steel bar of diameter ø20 mm with high-speed steel (HSS) tooling, the spindle speed would be calculated as:

$$\text{Rev/Min} = \frac{1000 \times S}{\pi \times d}$$

Where: S = Cutting speed of the work material (25 m/min)

d = Work diameter (20 mm)

π = 3.142 (Approximately 3)

$$\text{Rev/Min} = \frac{1000 \times 25}{3 \times 20}$$

$$= \mathbf{416\,rev/min}$$

Example 2: To **mill** high-carbon steel with a ø125 mm HSS cutter, the spindle speed would be calculated as:

$$\text{Rev/Min} = \frac{1000 \times S}{\pi \times d}$$

Where: S = Cutting speed of the work material (18 m/min)

d = Tool diameter (125 mm)

π = 3.142 (Approximately 3)

$$\text{Rev/Min} = \frac{1000 \times 18}{3 \times 125}$$

$$= \mathbf{48\,rev/min}$$

Appendix IX

Feed rates for milling

The feed rates given in this table are recommended feed rates for sharp HSS cutters working on robust machinery and using correct coolants/cutting fluids. They are correct for all materials. If in any doubt about feed rates, always use a slower one than calculated.

The suggested feed rates are based on $1 \times$ cutter diameter of axial cut depth and $\frac{1}{2}$ cutter diameter for radial depth of cut.

Feed rates for end mills and slot drills

Cutter diameter (mm)	Feed per tooth (mm) *for end mills*	Feed per tooth (mm) *for slot drills*
2	0.0075–0.013	0.013
3	0.013–0.025	0.018
6	0.025–0.037	0.025
12	0.037–0.050	0.050
25	0.05–0.064	0.090
50	0.064–0.075	0.13

Table feed per minute = feed per tooth × number of teeth on cutter × revolutions per minute of cutting tool

Feed/min = feed/tooth × no. of teeth × rev/min

The above feed rates should only be set if your cutting conditions are ideal (i.e. if you have sharp tools, cutting fluid available, secure clamping and a rigid machine tool.) If your conditions are not ideal, use a slower feed.

Example 1:
For end milling a brass workpiece with a HSS four tooth ø25 end mill, the feed rate would be found as follows:

(a) work out spindle speed:

$$\text{Rev/Min} = \frac{1000 \times S}{\pi \times d}$$
$$= 600 \text{ rev/min}$$

(b) calculate feed rate:

$$\begin{aligned}
\text{Feed/min} &= \text{feed/tooth} \times \text{no. of teeth} \\
&\quad \times \text{rev/min} \\
&= 0.05 \times 4 \times 600 \\
&= \mathbf{120\,mm/min}
\end{aligned}$$

Example 2:
For slot drilling a 12 mm wide slot in aluminium using a HSS slot drill, the feed rate would be found as follows:

(a) work out spindle speed:

$$\text{Rev/Min} = \frac{1000 \times S}{\pi \times d}$$
$$= 1666 \text{ rev/min}$$

(b) calculate feed rate:

$$\begin{aligned}
\text{Feed/min} &= \text{feed/tooth} \times \text{no. of teeth} \\
&\quad \times \text{rev/min} \\
&= 0.05 \times 2 \times 1666 \\
&= \mathbf{166.7\,mm/min}
\end{aligned}$$

Appendix X

Reaming allowances

Finished diameter (mm)	Drill size for machine reaming	Drill size for hand reaming
Up to 8	Hole size minus 0.25 mm	Hole size minus 0.2 mm
Over 8 up to 12	Hole size minus 0.3 mm	Hole size minus 0.25 mm
Over 12 up to 18	Hole size minus 0.4 mm	Hole size minus 0.35 mm
Over 18 up to 25	Hole size minus 0.6 mm	Hole size minus 0.5 mm
Over 25	Hole size minus 0.75 mm	Hole size minus 0.6 mm

Spindle speeds for reaming should be set at approximately half the equivalent drill speed.

Feed rates for reaming should be approximately twice the equivalent drill feed.

Tips on reaming

1. Ensure workpiece is held tightly.
2. Drill correct hole size before reaming.
3. Use cutting fluid or lubricant.
4. Ream straight after drilling and without moving the workpiece.
5. Select recommended speeds and feeds.
6. Look after the ream to maintain sharp cutting edges.
7. Continue to turn reamer forwards when withdrawing from workpiece.

Examples:

1. The pilot drill size for **hand reaming** a ⌀8 mm hole: $8 - 0.35 = 7.65$ mm
2. The pilot drill size for **machine reaming** a ⌀20 mm hole: $20 - 0.6 = 19.4$ mm

Torque wrench settings

The table below is guidance for torque wrench settings, use it only if manufacturer's recommended torque settings are not available.

Metric thread size	Torque wrench setting N/m		
	4.6 *quality*	8.8 *quality*	12.9 *quality*
M4 × 0.7	1.3	3.5	6
M5 × 0.8	2.7	7.1	12
M6 × 1	4.5	12.1	20.4
M8 × 1.25	11	29.4	49.6
M10 × 1.5	21.8	58.3	98
M12 × 1.75	38.1	102	171
M14 × 2	60.6	162	273
M16 × 2	95	252	426
M20 × 2.5	185	492	830

Imperial thread size	*Torque wrench setting lbf ft.*		
	'R' quality	*'S' quality*	*'X' quality*
$\frac{1}{4}$ Whit	8.2	9.1	15
$\frac{5}{16}$ Whit	17	19	31
$\frac{3}{8}$ Whit	30	33	55
$\frac{7}{16}$ Whit	48	53	88
$\frac{1}{2}$ Whit	70	79	130
$\frac{1}{4}$ BSF	9.1	9.1	17
$\frac{5}{16}$ BSF	18	19	33
$\frac{3}{8}$ BSF	32	33	59
$\frac{7}{16}$ BSF	51	53	95
$\frac{1}{2}$ BSF	77	82	143
$\frac{1}{4}$ UNC	8.2	10	15
$\frac{5}{16}$ UNC	17	20	31
$\frac{3}{8}$ UNC	30	36	55
$\frac{7}{16}$ UNC	48	58	89
$\frac{1}{2}$ UNC	73	86	136
$\frac{1}{4}$ UNF	9.3	10	17
$\frac{5}{16}$ UNF	19	21	34
$\frac{3}{8}$ UNF	34	36	63
$\frac{7}{16}$ UNF	53	60	99
$\frac{1}{2}$ UNF	82	91	152

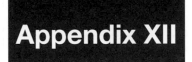 **Appendix XII**

Sample standard forms

If you do not use standard forms in your workplace, the forms below can be used. You must fill in the photocopies of these forms for all your personal records.

MATERIAL REQUISITION FORM

Material	Qty	Size	Job No.	Date
Requested by:			Approved by:	

TOOL REQUISITION FORM

Tool	Job No.	Loan date	Return date
Requested by:	Loan approved by:		

INSPECTION FORM

COMPONENT INSPECTION FORM			
Job title:			
Job number:	Date:		Checked by:
Component dimension:	*Limits of size:*	*Actual size:*	*Error:*
Recommendations/decisions:			

Appendix XIII

Some useful maths for engineers

Areas and volumes	Formula	Diagram
Area of rectangle	$A = l \times b$	
Area of triangle	$A = \frac{1}{2}b \times h$	
Area of circle	$A = \pi r^2$ or $A = \dfrac{\pi d^2}{4}$	
Circumference of circle	$C = 2\pi r$ or $C = \pi D$	
Volume of solid (with uniform cross section)	$V = $ area of end $\times$ height	

For example, volume of a cylinder $= \pi r^2 h$

Solving right angle triangles
Pythagoras' theorem $a^2 = b^2 + c^2$

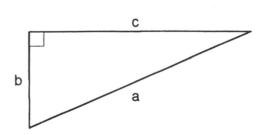

Trigonometry   Sine $S° = \dfrac{\text{Opposite}}{\text{Hypotenuse}}$

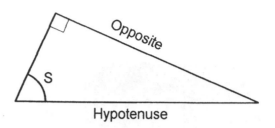

Cosine $C° = \dfrac{\text{Adjacent}}{\text{Hypotenuse}}$

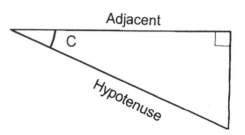

Tangent $T° = \dfrac{\text{Opposite}}{\text{Adjacent}}$

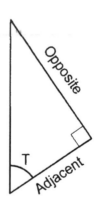

Percentages

For example, to add VAT @ $17\frac{1}{2}\%$

1. The cost of a micrometer is £36:50. What is the total price when VAT is included?

$$\text{VAT} = \frac{36.5 \times 17.5}{100}\%$$

Ratio

To cut a 1.5-mm bar into two pieces in the ratio of 2:3, add 2 and 3 = 5 and then divide 1.5 m by 5 = 30 mm. One piece will be 2 × 30 mm and the other 3 × 30 mm

Brass is 60% zinc and 40% copper.
Find the weight of zinc in 12 kg of copper, that is, ratio 6:4

Coolant must be 15% concentrate and 85% water. Ratio 15:85. 15 + 85 = 100, so divide the coolant volume by 100. Concentrate = 15× this volume and water is 85× this volume.

Density

$$\text{Density} = \frac{\text{Mass}}{\text{Volume}}$$

Appendix XIV

Risk assessment form

Risk assessment for:

Company:	Assessment undertaken:	Planned review:
Address:	By:	Time:
	Date:	
	Time:	Date:
Postcode:	Signed:	

Description of hazard	
Who is at risk	
Likelihood of accident	
Severity of accident	
Risk assessment result	
Actions:	1.
	2.
	3.
	4.
	5.
	6.
	7.
	8.

	9.
	10.
Special review notes:	

Appendix XV

Sample accident reporting form

If your company does not allow you to use a page from their accident book, you can use this sample form for Exercise 2:13. You must fill in this form and identify in Exercise 2.13 that you have done this.

Record of Accidents and Dangerous Incidents

Name of person injured Address:	Name of person making report

Date on which person was injured or when the dangerous incident occurred.	Time of accident or incident

Did the person require medical attention?	YES	NO
Did the person go to hospital?	YES	NO

Place where the accident / incident occurred

Please write a full account of how the accident/incident occurred:

Nature of injuries observed or reported

Action taken immediately and subsequently

Date: Signed:

                                   Print Name

Appendix XVI Work planning sheet

Procedure
1.
2.
3.
4.
5.
6.
7.
8.
9.
10.
11.
12.
13.
14.
15.

Tools and equipment selected

Index